TABLE OF CONTENTS

Page

List of Figures

List of Tables

AN EMPIRICAL ANALYSIS OF THE CASCADE SECRET KEY RECONCILIATION PROTOCOL FOR QUANTUM KEY DISTRIBUTION

I. Introduction

1.1 Problem Statement

The field of quantum key distribution is relatively young compared to other forms of cryptography. Although the security of different schemes of quantum key distribution have been studied, modified, improved upon, and even debunked, there is still a tremendous amount of work that needs to be accomplished to create a truly secure and reliable system. Improvements can be made at nearly every level from the actual hardware implementation and software protocols to the actual photon sources and sensitivity of detectors. With the introduction and sale of actual commercial quantum key distribution systems and hardware to be used for secure data transmissions, it is increasingly important that the specific and somewhat murky details of quantum key distribution systems are explored and completely characterized. Although quantum key distribution security is protected by the immutable laws of quantum mechanics, it is vital that current implementations are tested to account for the fact that technology has not reached the point where a true, simple and secure quantum key distribution system is feasible, lest we start protecting our confidential data with systems that may not be as robust and secure as we believe them to be.

One such component of a quantum key distribution system that requires more scrutiny is the error reconciliation protocol. Since the quantum channel of a key distribution system travels along a fairly lossy fiber optic cable or the even more lossy free space medium of the atmosphere, error correction and detection protocols are critical to the proper operation of quantum key distribution. Further, error detection is critical in determining if an eavesdropper is present in the system. This thesis focuses on the Cascade error reconciliation protocol. Although Cascade is one of the more heavily researched reconciliation protocols in the literature, there are still many questions that remain about the efficacy and efficiency of the protocol in quantum key distribution.

1.2 Research Purpose

The purpose of this research is to explore the Cascade reconciliation protocol in a simulation environment to better ascertain the boundaries of the algorithm's usefulness in the detection and correction of errors during the reconciliation phase of public channel quantum key distribution. This thesis seeks to answer the following research questions regarding the efficacy of Cascade:

- How does error sampling and error estimation impact the Cascade protocol?

- How sensitive is the Cascade protocol to incorrect error estimation?

- What is the ideal initial key length for the Cascade protocol?

- What is the relationship between the initial key length, error estimation, error rate, and information leakage in the Cascade protocol?

- What is the impact of burst errors on the operation of the Cascade protocol?

- What is the impact of an initial permutation of the key in the operation of the Cascade protocol?

- How does Cascade compare to other error reconciliation protocols?

1.3 Research Goals

The primary goal of this research is to determine the true bounds of the Cascade protocol in the reconciliation portion of quantum key distribution. The maximum and minimum failure points of Cascade are explored within multiple variable solution spaces and the effects of larger key sizes are investigated. In addition to answering the research questions and assessing the common metrics used to assess error reconciliation protocols, an exploration of the Cascade's performance with additional metrics is outlined in Chapter 3.

1.4 Thesis Structure

This research is organized into five chapters with the first chapter providing the introduction. Chapter 2 is a literature review of pertinent background information related to quantum key distribution and the Cascade reconciliation protocol. Chapter 3 discusses a methodology for conducting the experimental analysis of the Cascade protocol. Chapter 4 presents results from an empirical analysis of Cascade that is used to determine the usefulness and efficiency of the algorithm within the variable trade space. Chapter 5 provides the summary of the research findings and offers a proposal for future areas of interest in the domain of quantum key distribution.

II. Literature Review

2.1 Introduction

This chapter gives the reader an overview of concepts, definitions, and issues pertaining to the research. First, historical perspectives on cryptography and quantum key distribution are reviewed. Next, the evolution of quantum key distribution protocols is overviewed, followed by key reconciliation, error detection and correction, and privacy amplification. Then, the Cascade protocol is reviewed along with several other reconciliation protocols. Finally, this chapter concludes by discussing the importance of error detection and correction algorithms to quantum key distribution.

2.2 Cryptography

Cryptography is considered to be both a science and an art and examples range from the unusual arrangement of hieroglyphs on an Egyptian Tomb dating to around 2000 BCE to such modern day systems as the Advanced Encryption Standard (Singh, 2000). Cryptography has been around for thousands of years, and although the form of cryptography has changed over the years based on the technology of the times, the goal of cryptography has remained the same regardless of how it was implemented. Cryptography is the art of keeping secret information secret so that only intended recipients can share the information (Ferguson & Schneier, 2003). The actual information or plaintext consists of the plain message that the sender wishes to remain unknown to unauthorized recipients. The plaintext is encrypted into ciphertext using a cipher or algorithm to encode the information into an unreadable form. The intended recipient then decrypts the ciphertext back into plaintext using a key to extract the secret information.

Cryptography is used for four main purposes. The original and oldest purpose of cryptography was to maintain privacy between two or more parties. In essence, an unauthorized recipient of the ciphertext or an eavesdropper would not be able to make any sense of the ciphertext and be unable to extract the secret information encoded within. More recent uses of cryptography include authentication, integrity, and non-repudiation (Schneier, 1996). Authentication is the guarantee that the message is really from the entity that claims to have sent it. Integrity means that the encrypted message cannot be modified without the modification being detected by the intended recipient. Non-repudiation means that neither the sender nor the receiver can deny having sent or received the message respectively. With the advent of computing, cryptographic methods have increased in complexity both due to the ability of computers to handle increasingly complex ciphers and to the ability of using the same computers to break complex ciphers. The age of the computer in cryptology is indeed a two-edged sword.

Modern cryptographic methods utilize one or more of the following systems: symmetric or secret key, asymmetric or public key, and hash functions (Schneier, 1996). In secret key systems, both the sender and the receiver utilize an identical key to encrypt and decrypt the plaintext and ciphertext (Schneier, 1996). To ensure authentication and non-repudiation, each sender and recipient needs to have a unique set of keys for their transaction if there are more than two possible recipients or senders. As a result, the management of the keys can become complex and if there is a breach, new keys need to be sent to all recipients. This is known as key distribution. The problem with key distribution is that the keys must be securely distributed to all parties before encrypted messages can be sent. It is possible to overcome this vulnerability by having a trusted courier or agent deliver the key to the distant party, but it is easy to see how this can be a problem especially of one wants to key or rekey a satellite in orbit.

One solution that is used to counter a situation where the key needs to be shared combines the secret key system with another encryption method called asymmetric or public key (Schneier, 1996). Public key encryption pairs a public or known key and a secret key. If one of the keys (Public/Private) is used to encrypt the plaintext, the other key (Private/Public) from the same entity must be used to decrypt the ciphertext. The public key method can be more robust than symmetric key encryption in terms of key management, but it requires that the keys be generated in a way that prevents the reconstruction of one key from the other. One way hash functions are also used in modern cryptography systems. These hash functions take a longer length plaintext and transform it into a shorter ciphertext of fixed length (Schneier, 1996). Hash functions are designed so that the hash cannot be back calculated into the original plaintext.

For the purpose of outlining encrypted communication transactions, the literature commonly uses the names Alice, Bob, and Eve to describe the entities involved in cryptographic transactions. Alice and Bob are the senders and recipients of the information and Eve, for eavesdropper, is an unwanted and assumed to be malevolent third party that is trying to gain access to the plaintext (Ferguson & Schneier, 2003). Since most cryptographic protocols and methods rely on the computational difficulty of mathematical functions to maintain secrecy and security of the ciphertext, given enough computing power, most of not all protocols could be compromised. In addition, classical protocols do not function in a way that enables Alice and Bob to detect whether an eavesdropper, Eve, is potentially intercepting the ciphertext. In order to combat these shortcomings, a subset of quantum cryptography, quantum key distribution, is used to generate a secret, random key and share it securely between Alice and Bob, regardless of the existence of an Eve.

2.3 Quantum Key Distribution

Stephen Wiesner, a graduate student at Columbia University in the late 1960's and early 1970's, proposed the groundwork theory concerning the use of certain properties of quantum mechanics to allow for secure communications (Wiesner, 1983). Wiesner's work involved two ideas that utilized the properties of quantum mechanics. The first was a proposal to create fraud-proof bank notes by embedding a quantum particle within them that could not be duplicated by counterfeiters. Since the core idea of the theory required the ability to store the quantum particles for arbitrary periods of time, a feat technologically impossible at the time, Wiesner's work was not published and acknowledged until years later (Wiesner, 1983). The other idea concerned the transmission of two messages that were entangled so that reading one of the messages invariably destroys the other. Wiesner referred to this transmission as "multiplexing". By encoding the information in the messages into qubits of multiple conjugate bases, when one basis is observed or measured, the state of the qubits are collapsed onto that basis, destroying the information that was encoded within the other basis. Since the purpose of cryptography is to ensure that the information is not readable by an eavesdropper, the fact that in "multiplexing" an eavesdropper invariably destroys an intercepted message upon reading it was seen as an obtainable holy grail of cryptography and the field of quantum cryptography came into being (Brassard, 1993).

Quantum key distribution relies on the laws of quantum physics to maintain secrecy in key distribution as opposed to relying on the mathematically complex calculations that classical cryptography utilizes (Singh, 2000). Unlike classical information bits that can be only either a 1 or a 0, a quantum bit or qubit can be 1, 0, or a superposition of the two states. In quantum key distribution, the quantum states are represented as photons. These photons follow the tenants of quantum mechanics, specifically the Heisenberg Uncertainty Principle, the Principle of Indeterminacy, entanglement, and Schrödinger's Paradox (Townsend, 2000). Quantum key distribution also relies on the no cloning theorem, which states that an arbitrary and unknown quantum state cannot be copied identically. As a result, any measurement of a qubit will force it into a deterministic state that depends upon the method that was used to measure it. By collapsing the state in this fashion, the original state of the qubit is irrevocably lost or destroyed. Essentially, the qubit remains as a construct in 2-dimensional space until it is measured, collapsing the state onto a single one-dimensional axis pair. The term basis is used to describe the orthogonal axes that the qubit occupies when measured. It is in the basis that the qubit encodes information into the classical values of 0 and 1. In addition, by utilizing the fact that the polarization of a qubit encoded photon cannot be measured simultaneously in both the vertical/horizontal and diagonal bases, no measurement can determine the state of the photon qubit completely. Since the power of quantum key distribution is anchored in the fundamental laws of quantum dynamics, it is immune to the ever increasing computational power provided by the computer age if implemented correctly. The unique aspect of quantum key distribution is that an eavesdropper will, by simply observing the quantum channel, introduce changes in the information flowing through the channel in the form of errors. By monitoring the error rate of the quantum channel, the legitimate sender and receiver can determine if the channel has been compromised (Brassard, 1993). Unfortunately, since the state of the art in technology has not completely kept pace with the tenants of secure quantum key distribution, technological limitations force additional steps to be taken to ensure secure key distribution.

2.4 Evolution of QKD Protocols

Several quantum key distribution protocols have been created since Wiesner's pioneering work in quantum cryptography. The protocols are named by the authors' surname initials and year of publication. The first true quantum key distribution protocol was proposed by Charles Bennett and Gilles Brassard (Bennett & Brassard, 1984). Bennett and Brassard proposed a protocol, known as BB84, which took advantage of the multiplexing proposed by Wiesner. The basis of security for BB84, along with most QKD protocols, is that an eavesdropper will induce a measureable error in the quantum channel, allowing for their presence to be detectable by the sender and receiver. Figure 1 shows the actors and channels utilized by BB84 and most forms of QKD. Artur Ekert proposed E91, a protocol that distributes entangled photon pairs to each party in the transaction (Ekert, 1991). Like BB84, E91 utilized a classical channel and a quantum channel. SARG04, created by Scarani et al, is a protocol that directly modified the BB84 protocol. SARG04 replaces the single-photon source with attenuated laser pulses (Scarani, Acín, Ribordy, & Gisin, 2004). A more recent protocol, S09, was proposed by Eduin Serna (Serna, 2009). Unlike BB84 and its variants, S09 drops the classical channel and uses two quantum channels. As a result, Alice can transmit key material to Bob and other receivers securely without compromising any key material on a classical channel. Since BB84 is the most common protocol used in QKD systems, it will be the protocol discussed in detail below.

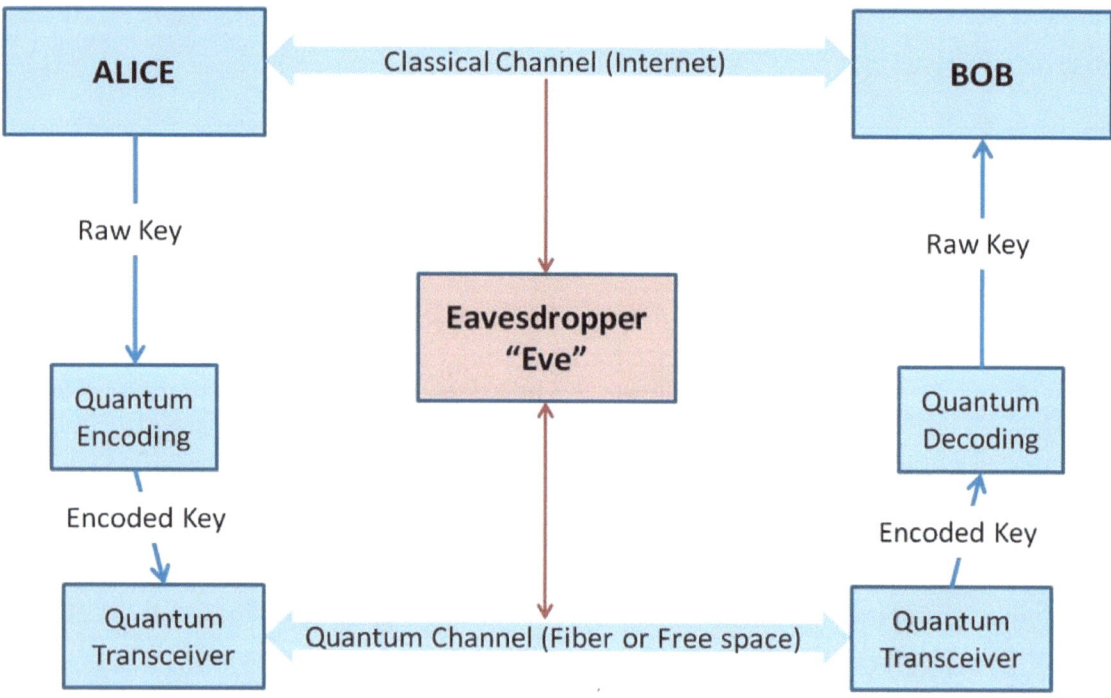

Figure 1. BB84 Actors and Channels

2.5 BB84 QKD Protocol in Depth

Relying on the security gained from its dependence on the no cloning theorem, BB84 utilizes photon polarization to encode classical bits into qubits. The protocol described by Bennett and Brassard utilizes a public classical channel and a quantum channel as seen in Figure 2. Since it is impossible to eavesdrop on the quantum channel without disturbing the information detectibly, the quantum channel is immune to passive eavesdropping. The opposite is true for the classical channel as long as it can be secured with classical authentication methods. Qubits are created by using photons polarized

using mutually unbiased polarization bases. There are several types of polarization that

can be chosen including rectilinear or 0 and 90 degree polarization, diagonal or 45 and

135 degree polarization, and circular or left-handed and right-handed polarization.

Illustrations of the various polarizations are shown in Figure 2. As long as the two bases

chosen are not orthogonal, they can be used to encode qubits securely since a qubit that is

polarized in one basis will effect a random measurement in the conjugate basis. By

choosing bases that are non-orthogonal, the protocol ensures that an incorrect

measurement of a qubit will destroy its encoded information. In addition, the destruction

of the information can be measured, revealing the presence of an eavesdropper.

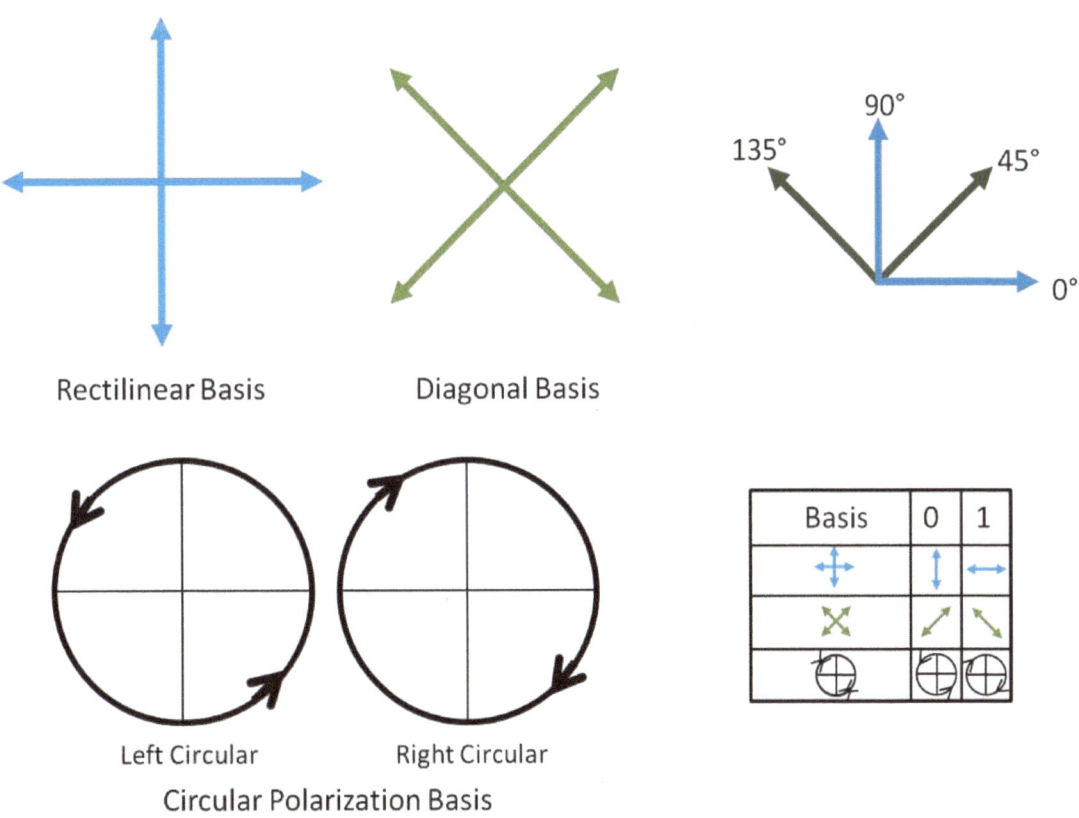

Figure 2. Possible BB84 Polarization States

BB84 utilizes the rectilinear and diagonal bases for encoding photons into qubits. The goal of BB84 is to satisfy the requirements of an encryption scheme known as One Time Pad (OTP). OTP is a completely secure encryption method if the key generation is truly random and the key is the same length as the message to be encrypted. In BB84, Alice, the sender, randomly generates photons in 0, 45, 90, or 135 degree polarizations. A critical assumption of the protocol is that the system only generates single photons and not bursts of photons. With today's state of the art in laser systems, it is not presently possible to construct single-photon sources or detectors. However, it is possible to modulate laser sources and bias detectors to increase the probability of single photon emission and detection while reducing the number of multi-photon packets that are emitted. In effect, the actual QKD systems have sources that are "mostly" single-photon. Alice generates the key material by randomly choosing key bits and basis. For BB84, a 0 is encoded as either a 90 degree polarized photon in the rectilinear basis or as a 45 degree polarized photon in the diagonal basis. Likewise, a 1 is encoded as a 0 degree polarization photon in the rectilinear basis or as a 135 degree polarized photon in the diagonal basis. Alice records the polarization state, basis and time that each photon is sent to Bob, and transmits the photons through the quantum channel one at a time. Bob, the receiver, receives the encoded photons and measures their polarization states, choosing his measurement basis randomly. Bob then records the measurement basis, the measured polarization state, and the time the photon was received. Since both the basis encoding and measurement is random, Bob can expect to correctly choose the right measurement basis 50% of the time. Once the entire key is transmitted, Alice and Bob utilize the classical channel to compare Alice's actual basis versus Bob's measured basis, discarding all improperly measured qubits for which the bases were mismatched. The key material remaining is called the sifted key. At this point, if there were no errors in transmission, Alice and Bob should have an identical, random key. The entire process is known as distillation. The steps of distillation are shown in Figure 3. Appendix A shows an example operation of the BB84 QKD protocol.

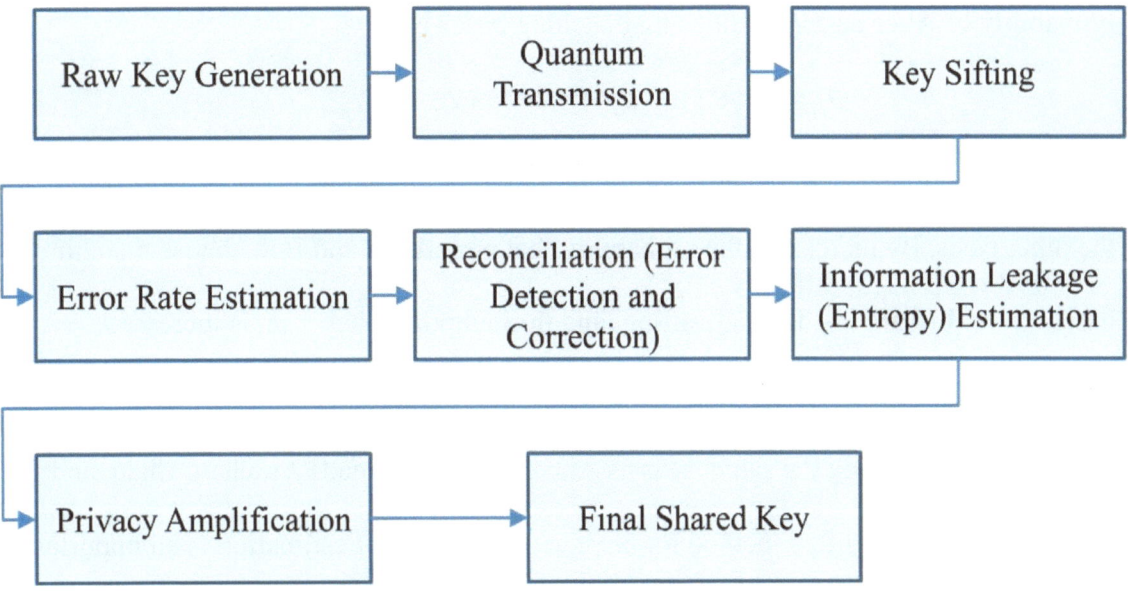

Figure 3. The Steps of Quantum Key Distribution

2.6 Key Sifting and Error Estimation

Since a fundamental aspect of quantum key distribution is the measurement of basis for bits sent and received, there will be bits in the raw key on which Alice and Bob disagree regarding basis. After Bob and Alice exchange raw key material, they must account for differences resulting from bits that Bob measured using a basis that did not match Alice's chosen sending basis. This process is known as sifting. During the sifting process, Bob transmits information over the public channel to Alice regarding the basis that he measured for each bit. Alice then responds with a yes for bases that she and Bob agreed upon. Once the mismatched bits are identified, Alice and Bob throw them out and the result is the sifted key. In 1999, Ardehali, Chau, and Lo published an optimization for the sifting process that can potentially result in increased throughput (Ardehali, Chau, & Lo, 1999). As opposed to an unbiased method of choosing and measuring bases, the authors recommend using a biased method for choosing the bases. Given that the probability of Alice and Bob choosing matching bases is

$$P = \sigma^2 + (1 - \sigma)^2 = 2\sigma^2 - 2\sigma + 1 \qquad (1)$$

where σ is the probability of choosing one basis and $1 - \sigma$ is the probability of choosing the other basis. By increasing the probability that both Alice and Bob choose matching bases, fewer bits are lost through sifting, and throughput, or key rate, is increased. Unfortunately, it is also possible for Eve to bias her measurement choice and increase her probability of choosing the same basis as Alice sends. This leads Ardehali, Chau, and Lo to also propose a modified method for error estimation. Error estimation is an important step of most quantum key distribution protocols. By choosing a set of random sifted bits and comparing them over the public channel, Alice and Bob can calculate an estimated error rate for the remaining bits. This is the error estimation method used by BB84. Alice and Bob use a predetermined "failure" error threshold to decide whether or not an eavesdropper is present. The actual error rate stems from both noise in the channel and possibly, interference from an eavesdropper. If the estimated error rate exceeds the acceptable error rate, then the key distribution process is terminated and started again from the beginning with a new raw key. In the literature, the most common failure error rate chosen is greater than or equal to 0.15 (Nakassis, Bienfang, & Williams, 2004). At 0.15 error rate, an eavesdropper could have intercepted over half of the bits transmitted. Ardehali et al. modify the error estimation process by choosing two samples of bits, one from each basis, and obtaining an error estimate for each. By comparing the two basis-biased error rates to the failure threshold and choosing the highest, Alice and Bob can effectively reduce any benefit that Eve may gain from biasing her basis measurements. Error estimation is also an important part of most error reconciliation protocols. If the estimated error rate is below the "failure" error threshold, Alice and Bob proceed with error reconciliation, as discussed below.

2.7 Reconciliation and Privacy Amplification

Error reconciliation is the process by which Alice and Bob correct errors resulting from noise and/or eavesdropping in the quantum channel. One of the most important goals of reconciliation is to enable Alice and Bob to detect and correct errors while simultaneously revealing as little information as possible to Eve. In fact, Eve gains as many bits of information concerning the secret key as the number of bits shared between Alice and Bob over the public channel (Van Assche, 2006). Christian Kollmitzer and Mario Pivk give the minimum information needed for Bob to correct his key as

$$H(A|B) = nH(p) \tag{2}$$

where n is the sifted key length, H is the Shannon entropy, A and B are Alice and Bob's sifted keys, and p is the error rate (Kollmitzer & Pivk, 2010). In the case of a discrete communications system, we use the binary entropy function described by MacKay (MacKay, 2005) for the Shannon entropy H seen in Equation 2:

$$H(X) = H_b(p) = -p \log_2 p - (1-p) \log_2(1-p) \qquad (3)$$

As a result, $n\mathrm{H}(p)$ is often used as a benchmark in reconciliation protocols as any

protocol reveals at least $n\mathrm{H}(p)$ bits as a lower bound. Thus, the theoretical limit of a

reconciliation protocol's efficiency η, is given as

$$\eta = 1 - H(p) \qquad (4)$$

(Yan, Ren, Peng, Lin, Jiang, & Liu, 2008). The closer a protocol is to the bound; the

more efficient it is in reducing leakage. In order to preserve the integrity and security of

their keys, Alice and Bob utilize one of two types of reconciliation. The first type is

interactive reconciliation, which consists of two-way interaction between Alice and Bob

over a public classical channel for the detection and correction of errors. The second type

of interaction is one-way reconciliation in which a decision is made beforehand regarding

how errors are detected and corrected. One-way protocols, by their nature, tend to reveal

less information over the public channel than interactive protocols where possibly many

messages are openly passed back and forth. Although there are many examples of

reconciliation protocols in both quantum key distribution and classical cryptography in

general, the focus of this paper will be the Cascade reconciliation protocol.

Privacy amplification is the step that follows reconciliation and involves the effort

to reduce the amount of knowledge concerning the secret key that Eve has gained during

reconciliation over the public channel and key transmission over the quantum channel.

The amplification step usually uses a hash function to create a new key. Since the final

privacy amplified key is generated from a much larger set of bits, each final key bit is

dependent upon multiple input bits, further reducing the amount of information that Eve

can have regarding the final key. Gilles Van Assche recommends that the reconciled key of length L be reduced by m bits and then further reduced by a variable parameter s, to be L-m-s, where m is the number of bits exposed during reconciliation. The end result is that Eve only has 2^{-s} bits of information regarding the key (Van Assche, 2006). The goal of the privacy amplification step is the output of a shorter key of which Eve has very little or no knowledge. Figure 4 shows the steps of distillation and the key length as the QKD process progresses.

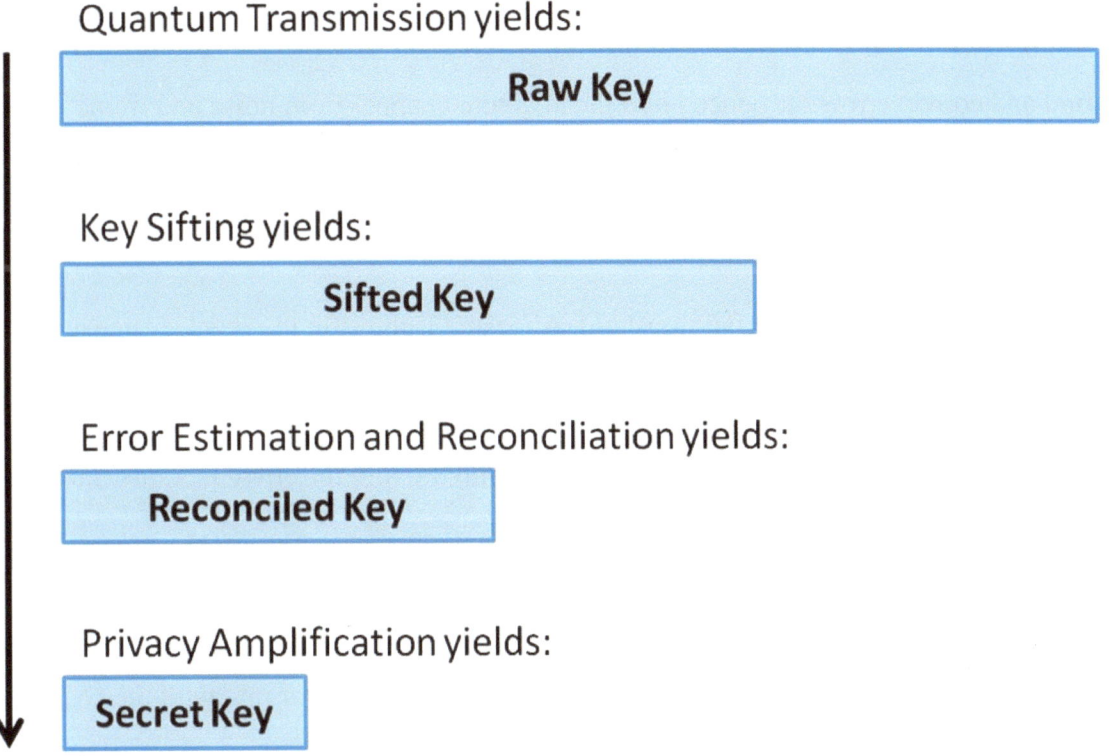

Figure 4. Distillation Process and Key Length

2.8 Cascade Reconciliation Protocol

The Cascade error reconciliation protocol is a modification of an earlier protocol from BBBSS, proposed by Bennett et al. in 1991 (Bennett, Bessette, Brassard, Salvail, & Smolin, 1992). BBBSS utilized an interactive error detection and correction protocol that the authors called Binary. BBBSS used a modified BB84 QKD protocol using a circular polarization basis instead of the diagonal basis. Other than the basis substitution, BBBSS operated as a BB84 implementation.

The Binary reconciliation protocol was used during the error reconciliation step to find and correct errors introduced into the quantum channel from noise and from eavesdroppers. After Alice and Bob obtain an error estimation based on a portion of their sifted key, they determine whether the error failure threshold has been breached. If the error rate is in excess of the fail threshold, Alice and Bob begin the raw key step again. If the estimated error rate is acceptable, Alice and Bob begin the first of a number of passes and use a predetermined random permutation, applying it to the sifted key bits. The purpose of the permutation is to attempt to spread out the error bits randomly and separate consecutive errors from each other. Alice and Bob then divide all of the sifted key bits into blocks of N_1 bits dependent upon the estimated error rate with the goal of having one or fewer errors remaining per block. Alice and Bob then use the classical channel to compare block parities. For blocks where the parities disagree, there must be an odd number of errors since an even number of errors would mask each other. The block is then divided in half into two smaller blocks of length $\frac{N_1}{2}$, and another parity check is conducted on the first sub-block. Since there is definitely at least one error in one of the sub-blocks, the parity of one sub-block reveals where the error has occurred. If the parity of the first sub-block passes, then the error is in the second sub-block. The sub-block with the error is further sub-divided and parity checked until the error bit is found. The error is then corrected.

The total number of parity bits exchanged is given as $\log_2(N_1)$, where N is the number of bits in the first block. Since the exchange of parity bits occurs on the classical channel, over which Eve can passively eavesdrop, it is assumed that the parity bits give Eve information about the secret key. By discarding the last bit of each block and sub-block involved in a parity check, the information that Eve gains about the key can be reduced. Since a number of even errors cannot be detected, the key is then permuted again and the Binary search check protocol is run again. In order to reduce the amount of discarded bits wasted when a parity check passes, essentially a lost potential key bit, the Binary protocol utilizes an additional process during later passes that wastes fewer bits. The authors refer to the new mode as confirm and bisect. In this stage, the parity of a random subset of the sifted key is calculated and compared. Subsets that fail the parity check are subdivided and checked using Binary. The authors of BBBSS recommend 20 random parity check passes, after which Alice and Bob assume that their reconciled keys are identical.

Brassard and Salvail created the Cascade reconciliation protocol as an improvement to Binary in terms of bits leaked during the reconciliation stage (Brassard & Salvail, 1994). By increasing the processing steps of the protocol, the authors claimed to have improved upon the information leakage problem and reduced bits leaked to the theoretical minimum needed to perform reconciliation. The operation of Cascade is nearly identical to Binary except that the error bit locations and values are retained for use in later passes. The error bit information is utilized in later passes so that the algorithm can correct all odd number errors and then cascade back through the previous passes to find even errors that were previously masked. A detailed description of the operation of Cascade is given below.

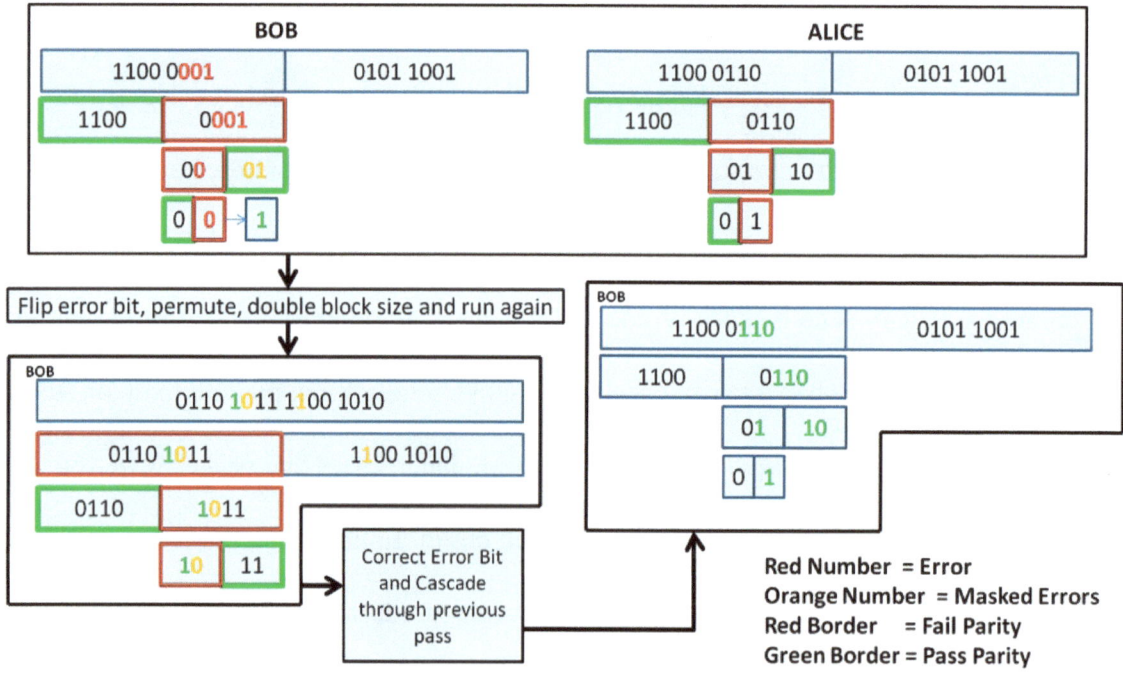

Figure 5. Binary Bisection and Cascade Operation

The Cascade error reconciliation protocol begins with an estimated error rate analysis of the sifted keys. The analysis is conducted as in BBBSS, with a subset of the sifted key compared across the classical channel between Alice and Bob. Alice and Bob then divide their sifted keys into blocks with a size dependent upon the error rate. The authors determine an ideal initial block size N_1 empirically as $0.73/p$ where p is the estimated error rate (Van Assche, 2006). If the error rate is 0.01, then $N_1 = 73$, or roughly 1 error per block (Brassard & Salvail, 1994). The block parities are then compared between Alice and Bob over the public channel, and the Binary protocol is utilized to correct errors. The final bit of each block involved is not discarded at this point. In addition, all of the information regarding error location is stored. After a permutation of the sifted key, a new pass is started. However, unlike in BBBSS, the block size is increased to $N_2 = 2N_1$ and another Binary search is conducted. Any errors found in this pass could only have resulted from two or more even number of errors that were masking each other in the previous pass. Using the information on error location stored from the previous pass, the Cascade algorithm returns to the shortest block that involved the initially corrected error from pass 1 and bisect it to find the hidden error. The protocol proceeds to operate in the same way with all discovered errors, cascading through previous passes to find and correct masked bits. After a number of passes, permutations, and cascades, the protocol finishes with low probability that errors still remain.

After the publication of Cascade, a number of works have involved attempts to improve upon the algorithm and reduce the information leaked over the public channel. Sugimoto and Yamazaki published an improvement that sought to reduce the number of leaked bits by noticing that Cascade is highly efficient at correcting errors in the first two passes, but with severely diminishing returns versus information leaked with remaining passes (Sugimoto & Yamazaki, 2000). Essentially, the authors were able to empirically determine that the first pass corrected, on average, 50% of the errors and the second pass corrected approximately 50% of the remaining errors. The cascade from the errors in pass 2 then corrected approximately 25% of the remaining errors. Instead of continuing the highly interactive and information leaking Cascade passes, the authors proposed that the third pass switch to the confirm and bisect method used in BBBSS to correct the remaining errors. This optimization allowed the modified Cascade algorithm to perform more closely to the theoretical output bound than the original Cascade protocol as fewer bits were leaked over the public channel in later passes.

Other modifications to Cascade involve the protocol's permutation step. The permutation before each pass is intended to spread errors out in order to reduce the chance that they will mask each other during parity checking. In 2000, Chen recommended replacing the random permutations with an interleaving technique optimized to reduce or eliminate error clusters from burst errors (Chen, 2000). Chen found empirically that information leakage could be reduced compared to standard Cascade. In 2002, Nguyen proposed another modification to the permutation method used in Cascade (Nguyen, 2002). Nguyen proposed creating a new block with the first bit from each previous block, a second block with the second bit from each previous block, and so on until each of the bits has been placed into new blocks. At this point Cascade is run normally. This method was shown to improve upon the operation of Cascade.

Nakassis et al. propose a complete modification of both the actual Cascade algorithm and the error reconciliation steps in general (Nakassis, Bienfang, & Williams, 2004). In their modification, the authors propose a new method that does not include the error estimation step that is usually conducted at the start of error reconciliation. Instead, the sifted key is divided and parities are checked until a specific number of sub-blocks are found that have parity errors. Using various calculations and empirical sampling of data, the authors created a table of possible number of altered bits using Bernoulli error distribution probabilities. By comparing the probabilities with the parity errors, the authors are able to determine probable remaining errors after each pass and utilize 7-bit hamming codes to correct errors when the remaining errors are estimated to be low enough. In 2006, it was shown that the theoretical upper bound for information leakage given for Cascade (and by inference, Binary) by Brassard and Salvail was possible to exceed in experimentation (Yamazaki, Nair, & Yuen, 2006). Rass and Kollmitzer attempted to improve the information leakage and efficiency of Cascade in 2009 by replacing Cascade's doubling block size with a dynamic method utilizing Bayesian statistics and information from previous executions to give the algorithm "learning capability" (Rass & Kollmitzer, 2009). In 2007, Capraro and Occhipinti attempted to improve the computational performance of Cascade by creating a parallel processing implementation that focused on maximizing key rate in a hardware setup (Capraro & Occhipinti, 2007). Boughattas, Iyed, & Rezig focused their efforts on the integration of privacy amplification within the Cascade protocol by studying "interactive discarding" methods to maximize both reconciliation efficiency and robustness against eavesdropping (Boughattas, Iyed, & Rezig, 2010).

2.9 Summary

This chapter summarized the evolution of quantum key distribution protocols and key reconciliation methods that are necessary to conduct the analysis presented in the thesis. The BB84 reconciliation protocol was outlined in detail and sifting, error estimation, reconciliation, and privacy amplification steps were discussed. An in-depth review of the Binary reconciliation protocol, and its successor Cascade, was performed to provide a basis of understanding for the research presented later in this paper.

III. Methodology

3.1 Introduction

The purpose of this chapter is to describe the environment in which the empirical analysis is conducted, explain the experimental design, and discuss the methods that were used to analyze the function and utility of the Cascade protocol. An overview of the Cascade software implementation is provided.

3.2 Cascade Implementation

The Cascade protocol is implemented in C++ code. Four classes are used to define and implement the Cascade Protocol: one class is used to store, track, and manipulate large bit buffers for Alice, Bob, and for general bit operations, one class is used to actually implement the Cascade protocol binary operations and cascading bit corrections during passes, a third class is used to provide an input framework for various modes available with the implementation and to drive the simulation, and the last class is used for permutation and reverse permutation of the sifted key between passes. For sifted key generation, the simulation utilizes the pseudo-random number generator Mersenne Twister developed by Matsumoto and Nishimura (Matsumoto & Nishimura, 1998).

The simulation is run using command line arguments in one of six available modes as detailed in Table 1. The mode selected determines which command line arguments are available for use. Input arguments are listed in Table 2 and include the following: sifted key length, maximum number of passes, permute first pass, number of trials, random number generator seed, optional Alice and Bob output file generation, pass zero block size, desired errors per block, initial error percent, error distribution type, sample rate, and sample method. For burst errors, the input arguments include number of bursts, burst length as a ratio of total errors, and burst distribution type. The output is printed to the screen with verbose data printed to a file. Output data includes number of bits before sampling, number of bits corrected, errors remaining after sampling, percentage of bits corrected, number of bits exposed, exposed percentage, reconciliation rate, estimated error rate, and error rate after sampling bits have been discarded.

MODE	ARGUMENTS
ALL	debug numbits maxpasses pfile permutepass0 trials seed
0	afile bfile Ko
1	afile bfile deb
2	errorrate errdist Ko
3	errorrate errdist deb
4	errorrate errdist samplerate sampletype deb
5	errorrate numbursts burstamount bursttype samplerate deb

Table 1. Cascade Simulation Modes and Arguments

ARGUMENT	DESCRIPTION (Usage)
debug	Enable verbose debug mode (0 or 1)
numbits	Number of bits in initial sifted key (10 to 1000000)
maxpasses	Maximum number of passes allowed for Cascade (2 to 10)
pfile	Enable optional custom permute file (filename)
permutepass0	Enable key permutation for pass 0 (0 or 1)
trials	Number of trials (1 to 1000000)
seed	Random number generator seed (0 to 65536)
afile	Alice custom sifted key buffer file (filename)
bfile	Bob custom sifted key buffer file (filename)
Ko	Pass 0 blocksize(1 to numbits-1)
deb	Number of desired errors per block (0.10 to 2.5)
errorrate	Initial actual error rate percentage in Bob's buffer (0.00 to 0.25)
errordist	Error distribution type (0=Uniform 1=Random 2=One Burst 3=Two Bursts 4=Three Bursts)
samplerate	Error sampling rate percentage (0.00 to 0.99)
sampletype	Error sampling method (0=uniform 1=random)
numbursts	Number of burst errors to introduce into Bob's buffer (1 to 10000)
burstamount	Percent of overall errors contained with bursts (0.00 to 0.99)
bursttype	Burst Error distribution type (0=uniform 1=random)

Table 2. Cascade Simulation Argument Descriptions

After arguments have been accepted and a trial has begun, the bit buffers are initialized and cleared and the permute arrays are prepared. A random sifted key is generated in Alice's buffer and copied to Bob's buffer. A corruption buffer representing bit errors introduced during the quantum channel transmission is created using the input regarding initial error rate. The corruption buffer is then used to corrupt Bob's buffer. At this point, Alice has a pure buffer with the original sifted key and Bob has a buffer with errors introduced. Randomly introduced errors are added using Bernoulli's Distribution where each bit has an equal probability of becoming corrupted. Single burst errors are introduced into the buffer in a block and the remaining errors are randomly distributed in the remaining buffer. Periodic burst errors are blocks of burst errors that occur at regular periods defined by the input number of bursts, burst length, and distribution type. At this point, the Cascade protocol begins. An error sampling array is initialized and a number of bits in Bob's corrupted buffer are sampled according to the input arguments. The sampled bits are compared to the same bits in Alice's clean buffer. From this comparison, an estimated error rate is computed and the sampled bits are discarded from both Alice and Bob's buffers. Using the estimated error rate, initial block size and number of blocks are computed using the input desired errors per block. Rules for minimum and maximum numbers of blocks and block size are enforced as per the original Cascade protocol. At this point, the Cascade simulation runs through a series of passes where Binary is used to dissect and compare bit parities for the blocks between Alice and Bob. When a parity mismatch occurs, Alice and Bob dissect the block and compare parities until the error is found. After correcting the error, the code "cascades" back through the previous block operations correcting error bits that may have been masked previously. The permute class code is then used to permute the buffer and Cascade proceeds to run through Binary again and correct more errors. When all errors are corrected or the maximum allowed passes is reached, the protocol stops. Summary statistics are printed to the screen and the data file. The protocol then proceeds to run through the next trial. After the trials are complete, overall trial run averages are output to the screen. Source code for the implementation of the Cascade simulation can be found on the included CD.

3.3 Simulation Metrics

The performance of an error detection and correction algorithm such as Cascade can be measured using a variety of methods. For the experiments outlined in this chapter, several important metrics are used to examine the Cascade protocol and determine its efficiency and effectiveness for reconciliation. Metrics utilized include estimated error rate, error estimation deviation, bits exposed, percent bits exposed, and throughput. By varying different Cascade parameters and plotting the various metrics, it is possible to gain insight into the impacts on the algorithm and determine how to maximize throughput and minimize information leakage, increasing the probability that the final key remains a secret shared only by Alice and Bob.

1. **Estimated Error Rate -** The estimated error rate is the estimated overall initial error rate that is computed by the algorithm during error sampling. The estimated error rate feeds directly into the next metric, estimated error deviation.

2. **Estimated Error Deviation-** Estimated error deviation is a metric that reveals how accurately the algorithm can estimate the rate of errors within the sifted key. Accurate error rate estimation is critical to the successful operation of Cascade.

Poor error rate estimation is revealed by a larger estimated error deviation as perfectly accurate error rate estimation will result in no error deviation.

3. **Bits Exposed-** Bits exposed or leaked is a measure of how many bits in the sifted key have been revealed over the classical channel, and potentially known to an eavesdropper. This metric is extremely important as bits leaked cannot be used for final key generation without increasing the possibility that the eavesdropper can guess the key. Since the purpose of QKD is to share secret key information between Alice and Bob, the ideal algorithm will leak as few bits as possible and allow more key bits to be usable.

4. **Percent Bits Exposed-** The percentage of bits exposed is the fraction of bits exposed over the entire number of bits. This metric is important in revealing the overall potential exposure of the protocol to an eavesdropper.

5. **Throughput-** Throughput is an important and widely used metric in determining the efficacy of reconciliation protocols. Throughput is defined as the number of available bits minus bits leaked and then divided by the total available bits. Since the bits sampled for error estimation are exposed as well, they must be taken into consideration when calculating throughput and subtracted as necessary. The throughput is therefore the percentage of reconciled bits remaining after error detection and correction. As a result, error detection and correction algorithms are ideally designed for maximizing throughput.

3.4 Simulation Environment

The simulations are run on a Windows 7 Ultimate desktop system running Ubuntu 10.10 (Linux Kernel 2.6.35-38) in a VMware Workstation 7.1.3 virtual appliance. The host computer has 12 GB of memory available and an Intel Core i7 960 Quad 3.20 GHz CPU. The virtualized guest operating system has 8 GB of memory and two dedicated processor cores. Simulations for the experiments are run from the command line with full parameterization. Status updates are printed to the screen and verbose statistical data are written to delimited files. Multi-parameter runs are completed using batch file processing and analysis and graphing of the data are conducted using Microsoft Excel.

3.5 Experiment 1: Error Sampling Rate

3.5.1 Objective

The intent of Experiment 1 is to discover the ideal error sampling rate for the protocol. Since the bits sampled for error estimation are discarded, the sampling rate chosen has a large impact on the bits remaining for the remainder of the algorithm and the length of key available for privacy amplification. As a result, it is beneficial to minimize the sampling rate as much as possible while ensuring that the effectiveness of error detection and correction is not diminished.

3.5.2 Simulation Parameters

Parameters that are affected by error sampling rate are estimated error rate, estimated error rate deviation, and bits exposed. Parameters held constant for this experiment are sifted key length, maximum number of passes (10), and desired errors per block (0.69 as suggested by Brassard and Salvail). The factors that are varied in this

29

experiment are error rate, which varies between 1%, 5% and 10%, and error sampling

rate, which is chosen to vary from 5% and 50% in 5% increments. The estimated error

rate deviation is computed as:

$$Estimated\ Error\ Deviation = \left| \frac{actual\ error\ rate - estimated\ error\ rate}{actual\ error\ rate} \right| \qquad (5)$$

3.5.3 Methodology

In this experiment, the ideal error sampling rate is determined by varying the error

sampling rate and the initial error rate in a series of 300 trials. The outputs of the trials are

graphed visually to determine how the various error sampling rates affect the estimated

error rate, estimated error rate deviation, and bits exposed. For determining success, the

smallest error sampling rate that gives the least variation within the estimated error rate

and bits exposed is compared to the smallest rate that gives the least estimated error rate

deviation. Successful trials also detect and correct all errors within the sifted key.

3.5.4 Assumptions and Limitations

Initial errors are assumed to be randomly distributed for this experiment. The

sifted key length is fixed at 500,000 bits and the maximum number of passes is set to 10.

The ideal number of errors for block, 0.69, was suggested by Brassard and Salvail;

however, they have reached this number empirically and no theoretical treatment has

been given.

Initial error rates greater than 10% are not considered as it is assumed that error

rates greater than 10% will not significantly impact the error estimation or deviation as

smaller initial error rates are expected to result in greater variance.

3.5.5 Expected Results

The effects of varying error sampling rates are measured by plotting error sampling rate versus estimated error rate, estimated error rate deviation, and percent of bits exposed. The results for three initial error rates are given in each plot. Results from varying the sampling rate are used to determine a suggested ideal sampling rate that has the least impact on the other factors. The suggested ideal sample rate is then utilized in the remaining experiments.

3.6 Experiment 2: Sifted Key Length, Error Estimation, and Bits Exposed

3.6.1 Objective

The intent of Experiment 2 is to discover the ideal sifted key length for the protocol. Since the sifted key length has a direct impact on the bits remaining for privacy amplification, it is important to optimize the sifted key length in order to account for bits lost to error sampling and bits exposed during parity checks in Cascade as well as reduce the processing time, increasing key rates.

3.6.2 Simulation Parameters

Parameters that are affected by sifted key length are estimated error rate deviation and bits exposed. Parameters held constant for this experiment are maximum number of passes (10), error sampling rate (.25) and desired errors per block 0.69. The factors that are varied in this experiment are error rate, which varies between 1%, 5%, 10%, and 15%, and sifted key length, which is chosen to vary from 4096, 10,000, 25000, 50000, 100000, 250000, 500000, 750000, and 1,000,000 bits.

3.6.3 Methodology

In this experiment, the ideal sifted key length is determined by varying the sifted key lengths and the initial error rate in a series of trials. 50 trials were conducted at each error rate and key length up to 500,000 bits and 20 trials were conducted at each error rate for key lengths greater than 500,000 for a total of 1200 trials. The results of the trials are graphed visually to determine how the varying sifted key length affects the estimated error rate deviation and bits exposed. For determining success, the error estimation deviation graph is analyzed and the smallest spread of deviation is chosen. For bits exposed, success is determined by finding the smallest variation of percent bits exposed for each initial error rate.

3.6.4 Assumptions and Limitations

Initial errors are assumed to be randomly distributed for this experiment. The error sampling rate is fixed at 25% and the maximum number of passes is set to 10. The number of trials for bit lengths greater than 500000 bits is set to 20 for each initial error rate as a single run at higher bit lengths takes significantly longer to process than lower key lengths. Also, initial error rates greater than 20% are not considered as the presence of error rates greater than 15% are assumed to be the result of an eavesdropper in the BB84 protocol and most other protocols (Nakassis, Bienfang, & Williams, 2004).

3.6.5 Expected Results

It is expected that a key length that is too small will result in a higher ratio of bits lost to bits remaining after the algorithm completes. Conversely, a key length that is too

32

large is expected to adversely affect algorithm performance with the parity checks within each pass increasing and as a result, the processing time is expected to increase as well. A conclusion is drawn about the necessary sifted key length with regards to the error estimation deviation and percent bits exposed. A proposal is made for an ideal sifted key length for the remaining experiments while keeping the processing time of the simulation in mind.

3.7 Experiment 3: Error Rate Estimation

3.7.1 Objective

The intent of Experiment 3 is to discover how well the error estimation of the algorithm performs. The accuracy of the error estimation has a direct impact on the overall performance of the detection and correction functions of Cascade. It is important that the calculated estimated error rate is as close as possible to the actual initial error rate to ensure that the algorithm creates an initial block size and number of blocks that satisfies the desired errors per block. Bob calculates the estimated error rate by sampling a percentage of the sifted key and comparing it over the public channel with the same subset of the original error-free sifted key held by Alice. The equation for calculating the estimated error rate is as follows:

$$Estimated\ Error\ Rate = \frac{number\ of\ errors\ detected}{number\ of\ bits\ sampled} \qquad (6)$$

3.7.2 Simulation Parameters

Parameters that affect the error rate estimation are the initial error rate, sifted key length, and error distribution. Parameters held constant for this experiment are maximum number of passes (10), error sampling rate (.25), sifted key length (500,000 bits) and desired errors per block (0.69). The factor in this experiment is the initial error rate, which varies from 1-20% in 1% increments.

3.7.3 Methodology

In this experiment, the error rate estimation is determined by varying the initial error rate in a series of trials. 10 trials were conducted at each error rate for a total of 400 trials. The results of the trials are graphed visually to determine how the varying initial error rate affects the estimated error rate. For determining success, the error estimation is graphed versus the initial error rate. A linear regression is conducted on the resultant line and a coefficient of determination is calculated.

3.7.4 Assumptions and Limitations

Initial errors are assumed to be randomly distributed for this experiment. A major limitation of the Cascade algorithm is that the error estimation depends heavily upon the error distribution and the method chosen for sampling. For this experiment, error sampling was conducted using random sampling as opposed to uniform sampling. This ensures that each bit within the sifted key has an equal chance of being sampled for error estimation.

3.7.5 Expected Results

The pairs of initial error rates and corresponding error rate estimations are plotted and the coefficient of determination is measured. Ideally, the data will be linear and the coefficient of determination of the linear regression should be very close to 1. The slope of the resulting regression the line should have a slope of 1 as well. The closer the coefficient is to 1, the more accurate the algorithm is in estimating the actual initial error rate and calculating block size and number of blocks accordingly.

3.8 Experiment 4: Burst Error Analysis

3.8.1 Objective

The intent of Experiment 4 is to discover how well the Cascade algorithm performs when the error distribution is bursty and not distributed randomly or normally. Although the literature describes Cascade as being more resistant to burst errors than other protocols such as Winnow, there has not been a decisive exploration of the limits of Cascade in detecting and correcting burst errors. Experiment 4 explores the bounds of Cascade in functioning in a burst error environment by looking at several different burst error distributions and scenarios. Types of burst error distributions explored are a single burst error in addition to a random distribution of errors, and periodic burst errors of varying lengths and frequency.

As can be seen in Figure 6, single burst errors occur as a cluster of errors within the sifted key that tends to be larger than the small clusters of errors expected in random error distributions. Periodic burst errors are two or more burst errors that occur with a periodic frequency within the key. One potential solution to burst errors is the introduction of a permutation of the sifted key before the first pass of Cascade is conducted. By permuting the key, burst error bits are spread out in a way that improves the operation of Cascade in a bursty environment.

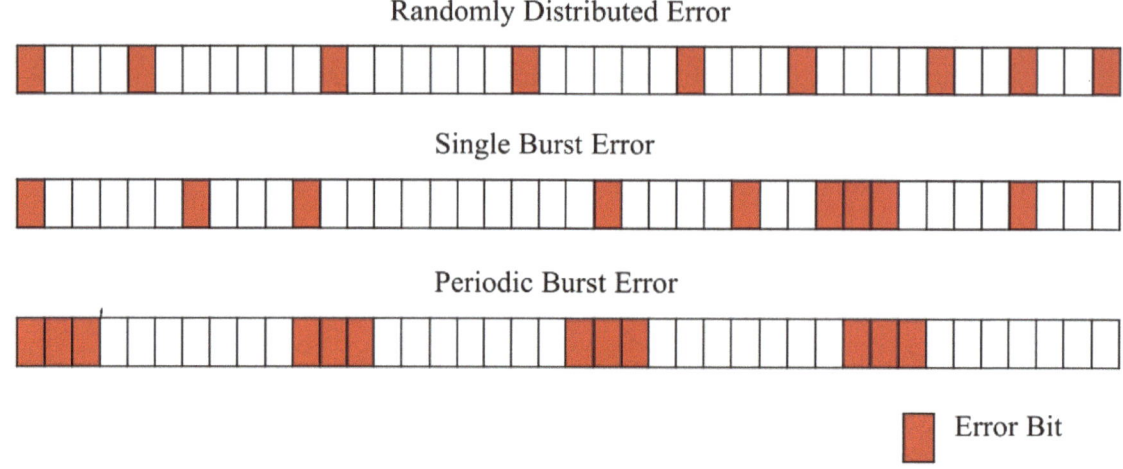

Figure 6. Error Distribution Types

3.8.2 Simulation Parameters

Parameters of this experiment include sifted key length, type of burst error distribution, burst error frequency, burst error length, and initial error rate. Parameters held constant for this experiment are maximum number of passes (10), desired errors per block 0.69, and sifted key length at 500,000 bits. The factors that are varied in this experiment are initial error rate, which varies between 0% and 18%, burst error distribution, burst error frequency, and burst error length. Burst error distribution is chosen to be uniform or random. Burst frequency is chosen to vary from 1 burst to 1000 bursts. The length of a burst error is given as a percentage of the total error bits that are contained within the burst error distribution and varies from 10%, 50% and 100%. As a result, the percentage of errors contained within the bursts is proportional to the overall error rate and error bits not contained with the burst are randomly distributed within the rest of the key. Trials are run both with and without an initial permutation. For permuting the first pass, a permute function is used to shuffle the bits of the initial key before the first pass of Cascade is run.

36

3.8.3 Methodology

In this experiment, the ability of Cascade in handling burst errors is determined by varying the sifted key lengths and the initial error rate in a series of trials. 10 trials are conducted at each error rate from 1% to 18% for a series of burst error distributions, frequencies, and lengths. Each run is accomplished with and without a permutation before the first pass. The results of the trials are graphed visually to determine how the varying burst error distribution types, frequencies, lengths, and error rates affect the operation of Cascade with and without a permuted first pass. The metric chosen to represent the Cascade operation is throughput. Both the burst error throughput and throughput without burst errors is presented with permute and no permute data represented.

3.8.4 Assumptions and Limitations

The percentage of errors not contained with the burst clusters are assumed to be randomly distributed for this experiment. It is also assumed that the burst errors are contained within solid clusters and that the frequency of bursts is constant. A major limitation of the Cascade algorithm is that the error estimation depends heavily upon the error distribution and the method chosen for sampling. For this experiment, error sampling is conducted using random sampling as opposed to uniform sampling. This ensures that each bit within the sifted key has an equal chance of being sampled for error estimation. A major limitation is the strength of the permutation function chosen to permute the bits of the sifted key. Certain permutations are expected to be more optimal in reducing error clusters than other permutations; however, only one permutation function is utilized in this experiment and it is by no means expected or intended to be an optimal choice. The exploration to find and define optimal or more optimal permutation functions is left for future studies.

3.8.5 Expected Results

It is expected that the successful operation of Cascade will be dependent upon the presence and frequency of burst errors as the error estimation will be directly impacted by burst error length and location. As a result, the Cascade calculation for block size will be impacted and the actual errors per block may differ from the desired errors per block. The throughput of Cascade will be negatively impacted by burst errors. By applying a random permutation before the first pass of Cascade, the effect of burst errors on throughput is expected to be minimized.

3.9 Summary

This chapter presented the research methodology for this study. The conceptual framework was defined and the parameters, factors, experimental designs, and simulation environment were summarized. The data collection procedures and data analysis methods were also discussed. Finally, this chapter reviewed the empirical analysis that will be presented in the following chapter.

IV. Results and Discussion

4.1 Introduction

The results of the experiments described in the Methodology Section are provided in this chapter. An analysis of the experimental results is presented and key findings are highlighted. In addition, an overview of the performance of Cascade is provided.

4.2 Experiment 1: Error Sampling Rate

The plot shown in Figure 7 shows the data collected in the error sampling rate experiment for error sampling rates from 5% to 50% and initial error rates of 1%, 5%, and 10%. The x-axis is the error sampling rate, and the y-axis is the estimated error rate. The experimental runs shown in Figure 7 were collected from 300 simulation trials.

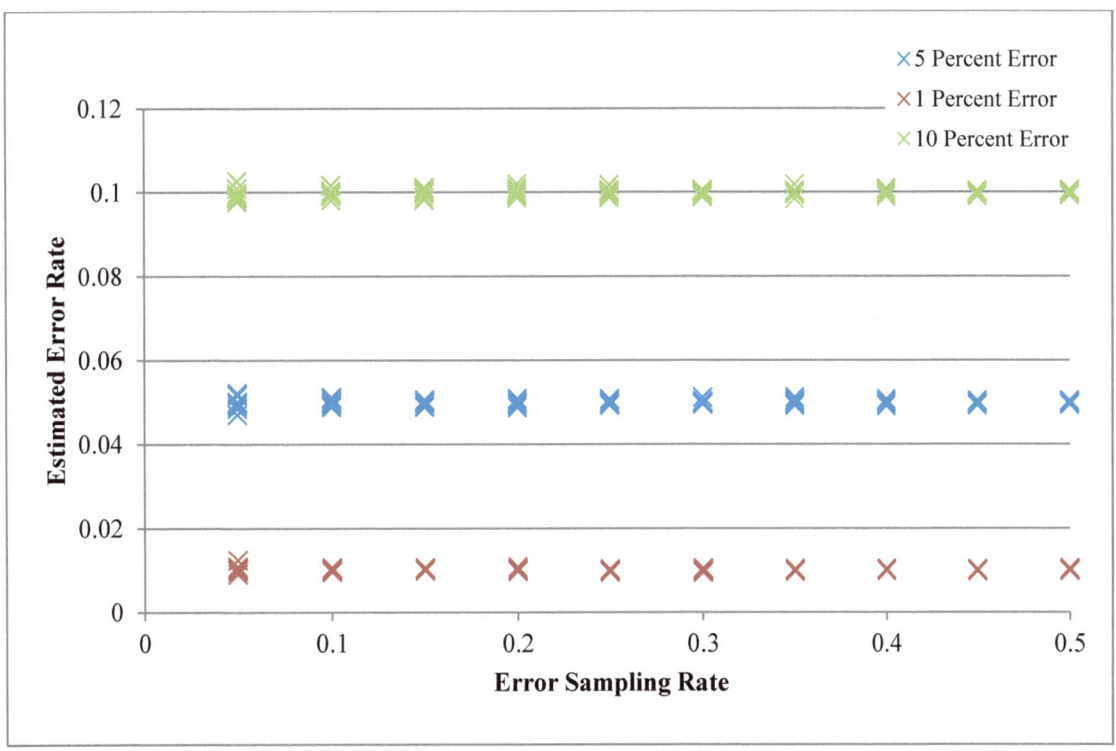

Figure 7. Error Estimation as a Function of Error Sampling Rate

From Figure 7, it can be seen that the error sampling rate chosen clearly has an effect on the error estimation rate. As the error sampling rate approaches zero, the estimated error rate begins to deviate from the actual error rate depending upon the magnitude of the initial error rate. For initial error rates of 1%, 5% sampling was measured to be nearly 3.5 times the deviation seen at 10% sampling. Similarly, for initial error rates of 10%, the 5% sampling spread was 1.36 times greater than the 10% sampling. Deviation is seen at all initial error rates. In addition, the higher the error sampling rate, the more likely the estimated error rate is to accurately reflect the actual error rate. Another revelation is that an increased initial error rate results in more fluctuation and uncertainty in the estimated error rate. Since the bits sampled for error estimation are discarded and unusable in creating a final key, it is important that a balance is maintained between choosing enough sample bits to accurately estimate error rate and retaining enough of the sifted key to perform adequate privacy amplification.

Figure 8 shows a plot of error sampling rate versus the estimated error rate deviation given in Equation 5. As before, the data is for error sampling rates between 5% and 50% and initial error rates of 1%, 5%, and 10%. The y-axis is the estimated error rate deviation, or how much the estimated error rate deviates from the actual initial error rate. Interestingly, the error rate deviation is the highest for the lowest initial error rate of 1%, approaching nearly 10% deviation from the actual initial error rate. An initial error rate of 10% yields a much lower error rate deviation, with the highest deviation measured at only 2.5% from baseline. There is also clearly a relationship between the error sampling rate and the error rate deviation in that more bits sampled for error estimation yields a more accurate estimation of the error rate. This trend is seen for all initial error rates and the measured maximum error rate deviation for all initial error rates up to 10% was no more than 3.8% down to error sampling rates of 35%.

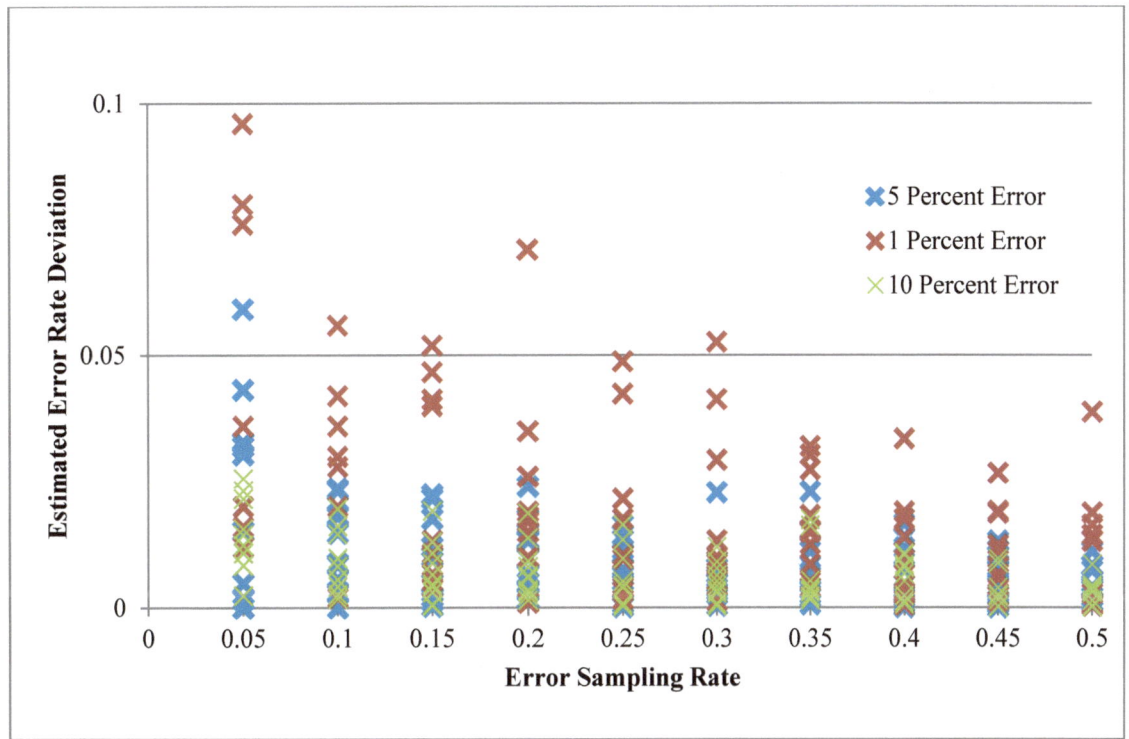

Figure 8. Estimated Error Rate Deviation as a Function of Error Sampling Rate

Figure 9 shows the percentage of bits exposed for error sampling rates from 5% to 50% and initial error rates of 1%, 5%, and 10%. The percentage bits exposed is relatively stable for all measured error rates as can be seen in Figure 9, and varying the error sampling rate seems to have no impact. Although the error sampling rate chosen has a direct effect on the overall bits exposed during the operation of the protocol due to the bits sampled being compared over the public channel, there does not appear to be an effect on bits exposed after the rate has been calculated as the bits sampled for error estimation are discarded before the error detection and correction phase. Consequently, the chosen sampling rate does not impact the final bits exposed in the corrected key used for privacy amplification.

As a result of the findings in Experiment 1, an error sampling rate of 25% is suggested to maintain a balance between error rate estimation and usable sifted key bits. An error sampling rate of 25% yields a maximum deviation of less than 2% for an initial error rate of 10% and less than 5% deviation for an initial error rate of 1%. An error sampling rate of 25% is satisfactory for use in the remaining experiments.

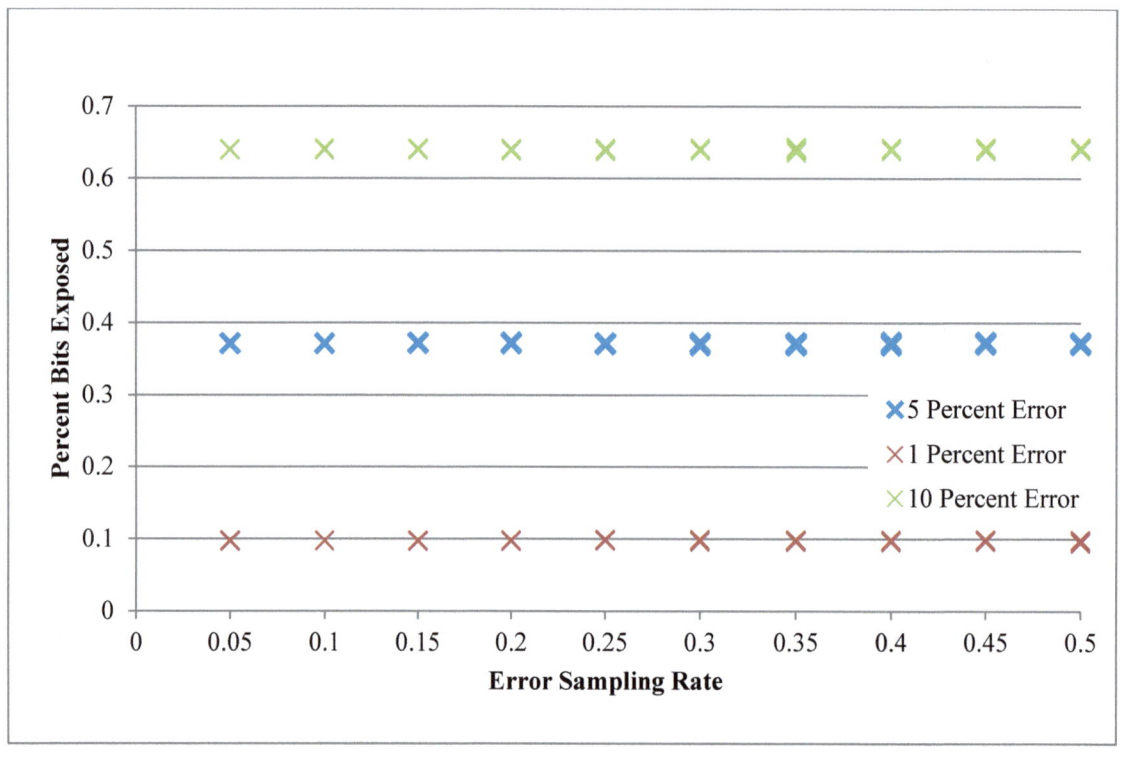

Figure 9. Exposed Bits as a Function of Error Sampling Rate

4.3 Experiment 2: Sifted Key Length, Error Estimation, and Bits Exposed

The data from experiment 2 was collected using sifted key lengths varying from 50,000 bits to 1,000,000 bits and initial error rates of 1%, 5%, 10%, 15%, and 20%. The plots shown in Figures 10 and 11 show the x-axis as sifted key length and the y-axis as the error estimation deviation. The plots for experiment 2 represent 1200 trials.

Figure 10 shows initial error rates of 1% to 10% and Figure 11 shows initial error rates from 10% to 20%. It can be seen that error rate deviation is inversely proportional to sifted key length since deviation increases as sifted key length decreases. The largest increase in error estimation deviation is in the smallest initial error rate tested, 1%. The deviation approached nearly 16% for a 1% initial error rate with 50,000 bits of sifted key. For the same sifted key length, the deviation was not more than 3.5% for a 20% initial error rate. For key lengths of 500,000 bits, the maximum deviation was only 4.8% at 1% error rate and less than 1.1% for a 20% initial error rate. For sifted key lengths of 750,000 bits and greater, the improvements in error estimation deviation appear to plateau at around 2.5% for 1% initial error and less than 1% deviation at 20% initial error rates.

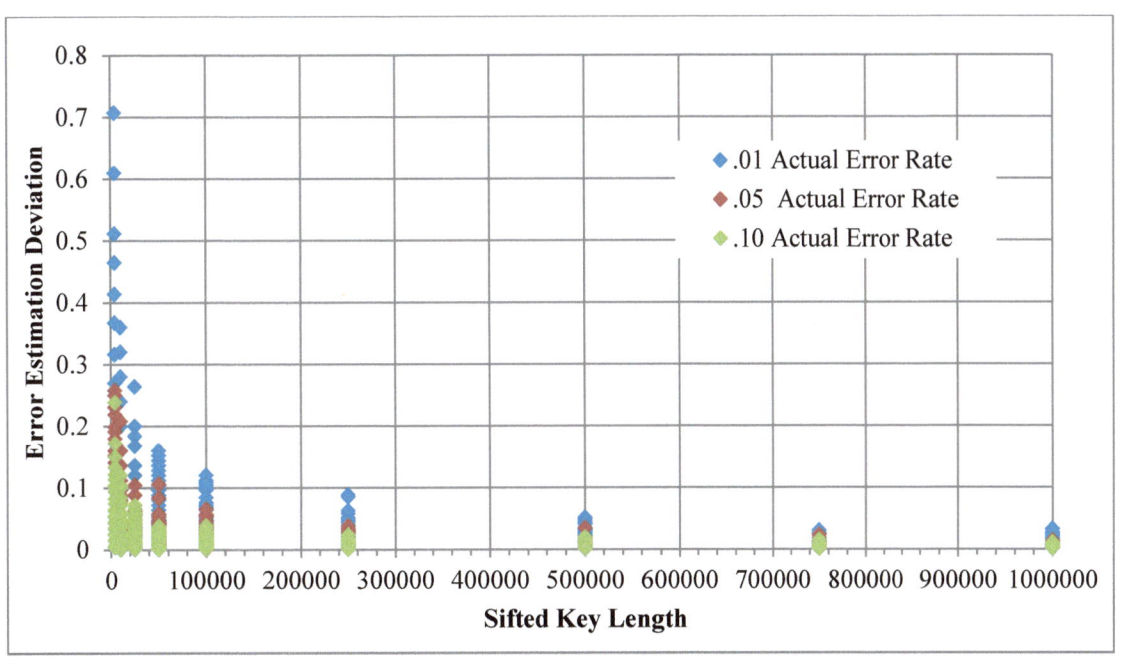

Figure 10. Error Estimation Deviation as a Function of Key Length (1-10%)

43

From this data, it appears that shorter sifted keys result in increased error estimation deviation especially at lower initial error rates. This will force the Cascade protocol to over or underestimate the number of blocks and therefore the block size calculation during the Binary operations. Since the block size is typically doubled after each pass, the error will be doubled as well, resulting in more passes required to detect and correct error bits.

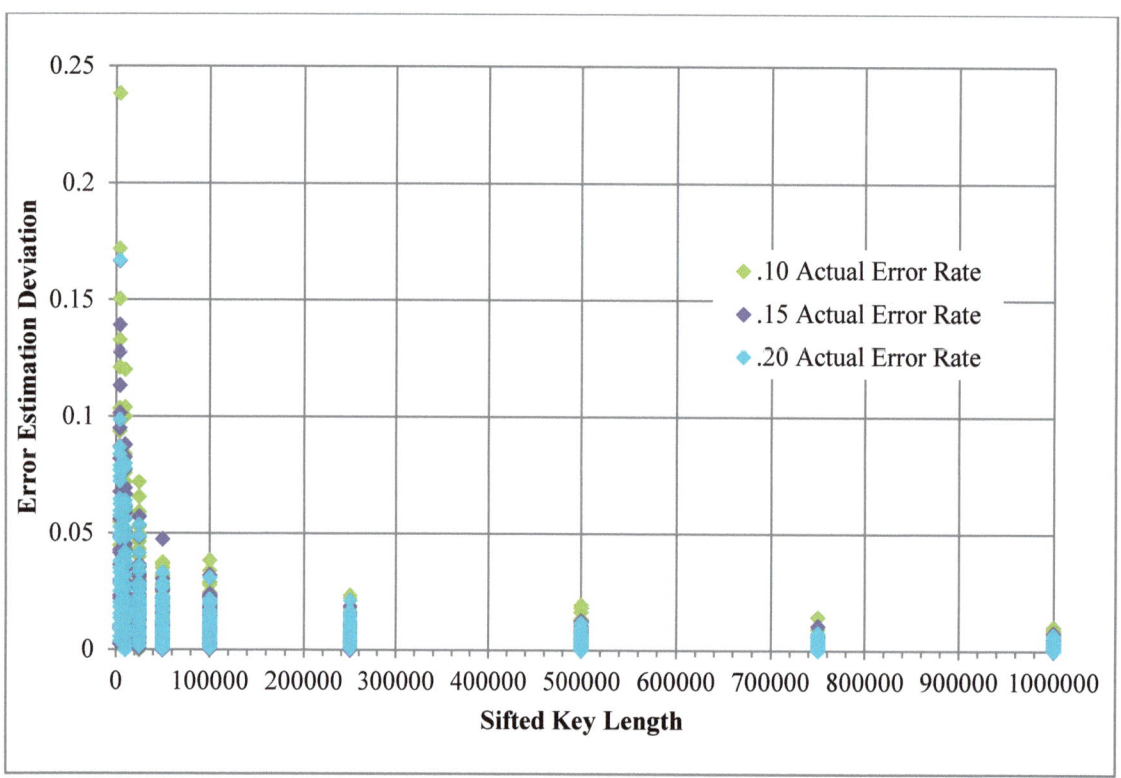

Figure 11. Error Estimation Deviation as a Function of Key Length (10-20%)

In Figures 12, 13, 14, and 15 the sifted key length is represented on the x-axis and the percentage of bits exposed is represented on the y-axis. Figure 12 is bits exposed at 1% initial error rate, Figure 13 is 5% initial error rate, Figure 14 is 10% initial error rate, and Figure 15 represents 15% initial error rate. From these figures, it can be seen that the size of the sifted key length has a direct impact on the percentage of bits exposed during the Cascade Protocol operation.

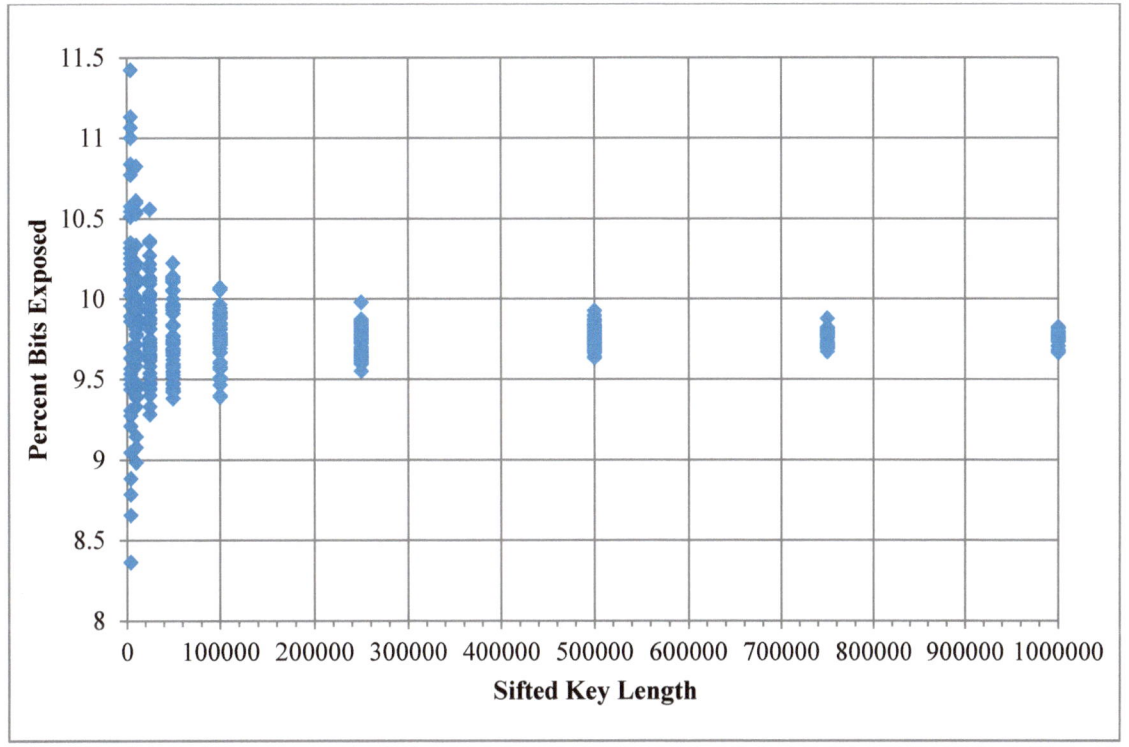

Figure 12. Bits Exposed as a Function of Key Length (.01 Actual Error Rate)

For 1% initial error rate, shown in Figure 12, the variation between percentages of bits exposed is greatest for the smallest sifted key length of 4096 bits. The variation is reduced as sifted key length increases, and appears to reach a minimum variation at around 750,000 bits and larger. The same affect can be seen in Figures 13, 14, and 15. Interestingly, the maximum percentage of bits exposed in the data for each key length and error rate is reduced as key length increases. The minimum percentage of bits exposed increases as key length increases. However, the trend of both maximum and minimum bits exposed appears to be asymptotic in nature, leveling out as the sifted key length increases. This asymptotic phenomenon appears at each initial error rate and is

45

evident in Figures 13, 14, and 15 as well. It is evident that the variance between bits

exposed for all error rates is ultimately dependent upon the sifted key length chosen.

However, the difference in bits exposed for 1% initial error rate is only 0.2% for keys

greater than 750,000 bits. Although increasing the sifted key length reduces the spread of

bits exposed, it also increases the processing time of the algorithm as well as the time

needed to transmit and receive the bits. An interesting aspect of the data plotted in

Figures 12, 13, 14, and 15 can be seen in the distribution of data points for each key

length. There appears to be a shift in the 5% actual error rate plot where the center

groupings of percent bits exposed for key lengths greater than 100,000 bits are translated

upwards compared to the center of the lower key length points. It appears that the

maximum information leakage at 5% error rate tends to be increased slightly and that

there is not as high of an improvement in leakage as the key length is increased compared

to the other measured error rates. In addition, the differences between the maximum and

minimum bit leakage is greater for 5% error rates as the variation for 5% error rate is

1.5% compared to the other rates that have a maximum bit leakage variation of only 1%.

The variance in bit leakage for 5% error rate does not approach the variances of the 1%

and 10% error rates until key lengths of 750,000 bits are utilized.

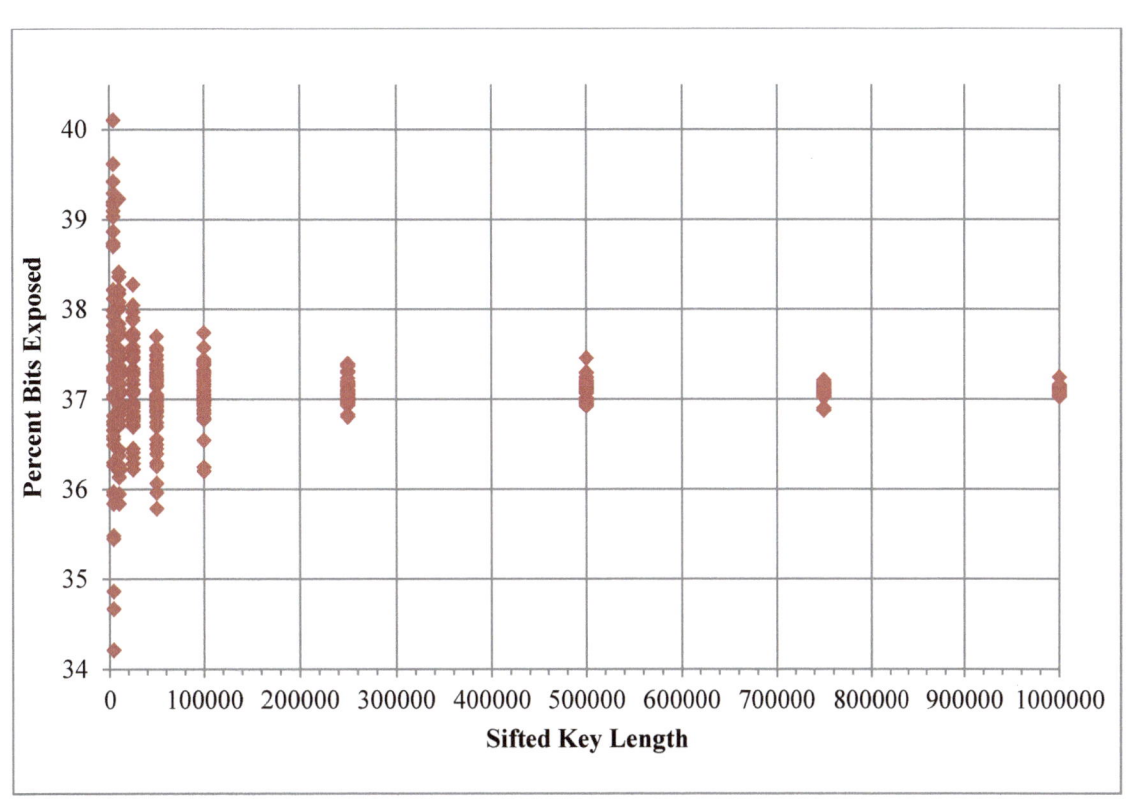

Figure 13. Bits Exposed as a Function of Key Length (.05 Actual Error Rate)

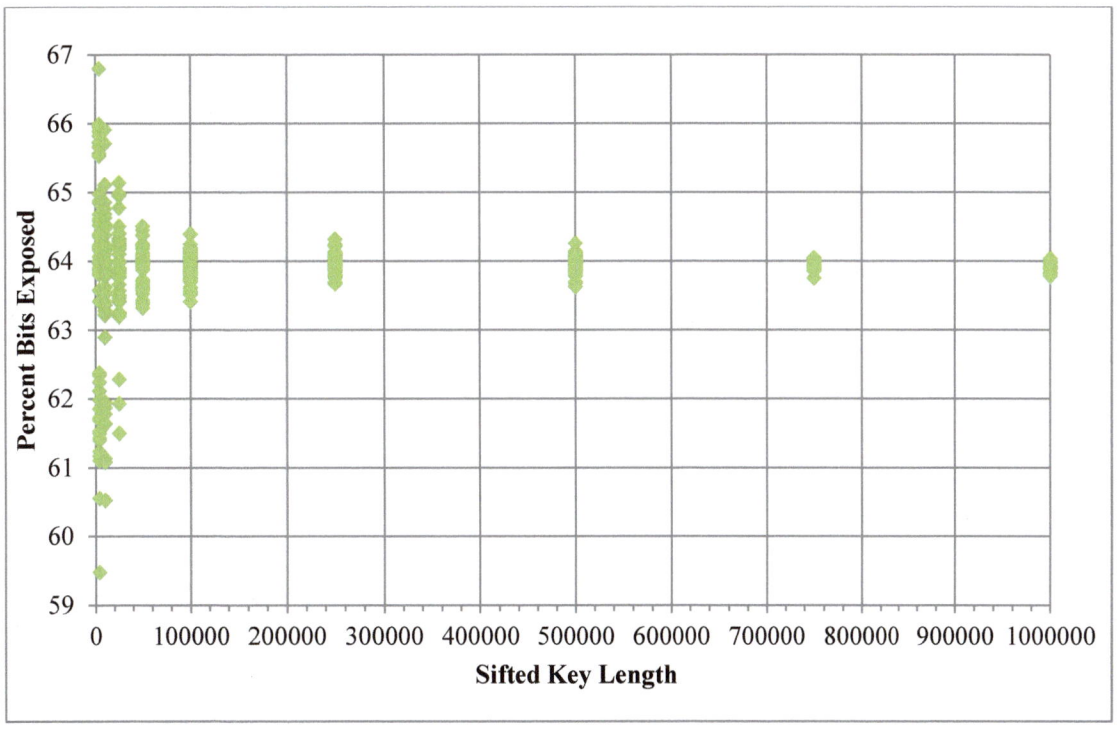

Figure 14. Bits Exposed as a Function of Key Length (.10 Actual Error Rate)

As a result of the findings in Experiment 2, a minimum sifted key length of 500,000 bits is suggested to maintain consistently low bits exposed and reduce estimated error deviation. This will ensure that the calculated block size and number of blocks remains closer to maintaining the desired errors per block. The number of bits exposed is minimized accordingly. A sifted key length of 500,000 bits yields a maximum deviation of less than 6% for initial error rates of up to 20% and significantly less variation in the percentage of bits exposed during the operation of Cascade. As the time required to process the sifted key in the simulation is proportional to the sifted key length, a sifted key length of 500,000 bits was used in the remaining experiments to reduce simulation runtimes to manageable durations while ensuring experimental validity.

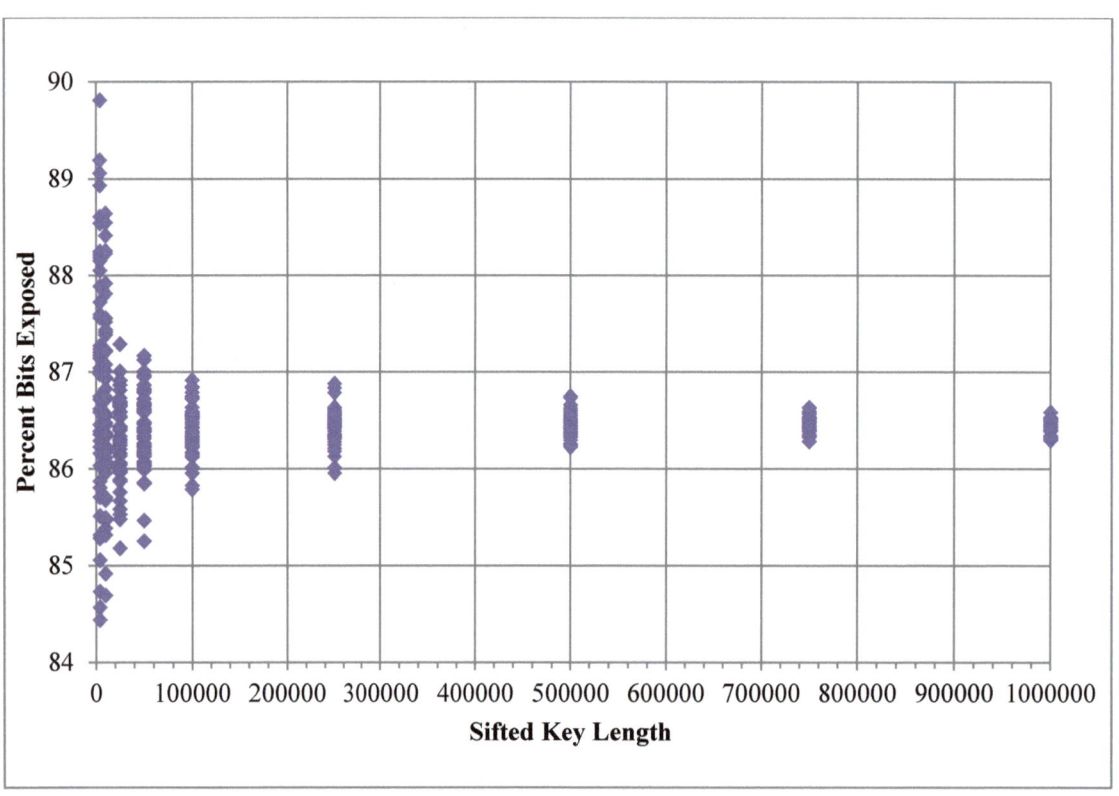

Figure 15. Bits Exposed as a Function of Key Length (.15 Actual Error Rate)

A comparison of bits exposed found in other implementations of Cascade in existing research can be seen in Table 3, below. Considering that Yamazaki, Nair, and Yuen used 10,000 bits for their sifted key, the exposed bit percentages actually compare quite well at low error rates. The data collected at 3% error rate also corresponds well with Pearson's results. The significant differences at higher initial error rates between Yamazaki, Nair, and Yuen's data and the author's data is most likely due to differences in the implementation of the Cascade protocol and experimental setup.

Initial Error Rate	Author (500K bits)	Author (10K bits)	*YNY (10K bits)	Author (4096 bits)	**PER (4096 bits)
0.01	0.0979	0.0982	0.0906	0.1046	----------------
0.03	0.2464	0.2468	------------	0.2465	0.2338
0.05	0.3710	0.3721	0.3382	0.3785	----------------
0.10	0.6394	0.6355	0.5753	0.6375	----------------
0.15	0.8646	0.8653	0.7678	0.8711	----------------

* (Yamazaki, Nair, & Yuen, 2006) ** (Pearson, 2004)

Table 3. Percent Bits Exposed Comparison to Prior Research

4.4 Experiment 3: Error Rate Estimation

The plot shown in Figure 16 shows the data collected in the error rate estimation experiment for initial error rates from 1% to 20% and a sifted key length of 500,000 bits. The x-axis is the initial error rate and the y-axis is the estimated error rate. The plot for experiment 3 is for 400 trials.

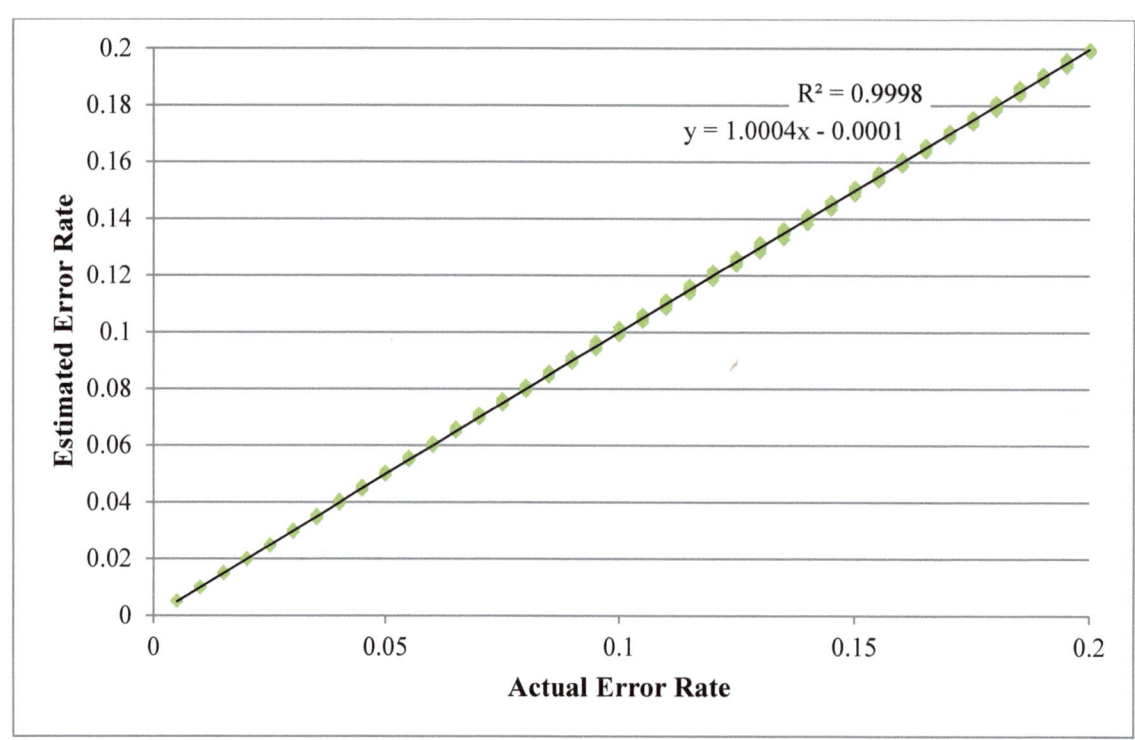

Figure 16. Actual Error Rate versus Estimated Error Rate

The pairs of initial error rates and corresponding error rate estimations were plotted and the coefficient of determination was measured to be 0.9998. As expected, the relationship is linear. The slope of the resulting regression line is nearly 1 as well. The Cascade protocol is very accurate in estimating the actual initial error rate. As a result, the protocol should correctly calculate block size and number of blocks accordingly. The error estimation for the algorithm is accurate for all initial error rates from 1% to 20%. It can also be seen in the plot that, although the estimated error rate data points deviate more as the actual error rate increase, the estimations are still very well grouped even at high error rates. Cascade appears to not only be able to accurately estimate error rates, but it can consistently estimate error rates accurately as well.

4.5 Experiment 4: Burst Error Analysis

The plot shown in Figure 17 shows the data collected for the burst error experiment for a single burst error containing 50% of the error bits and initial error rates of 1% to 18%. The x-axis is the chosen error sampling rate, and the y-axis is the estimated error rate. The plots for experiment 4 represent 1200 trials. For error rates above 18%, the algorithm exposes 100% of the sifted key bits. Cascade is more robust against burst errors than initially expected: this is most likely the result of the already extensive binary parity checks that are performed during the passes. Plot 17 shows Cascade's throughput for a sifted key with random error distribution, a single burst error with an error ratio of 50%, and a single burst error with permutation before the first pass.

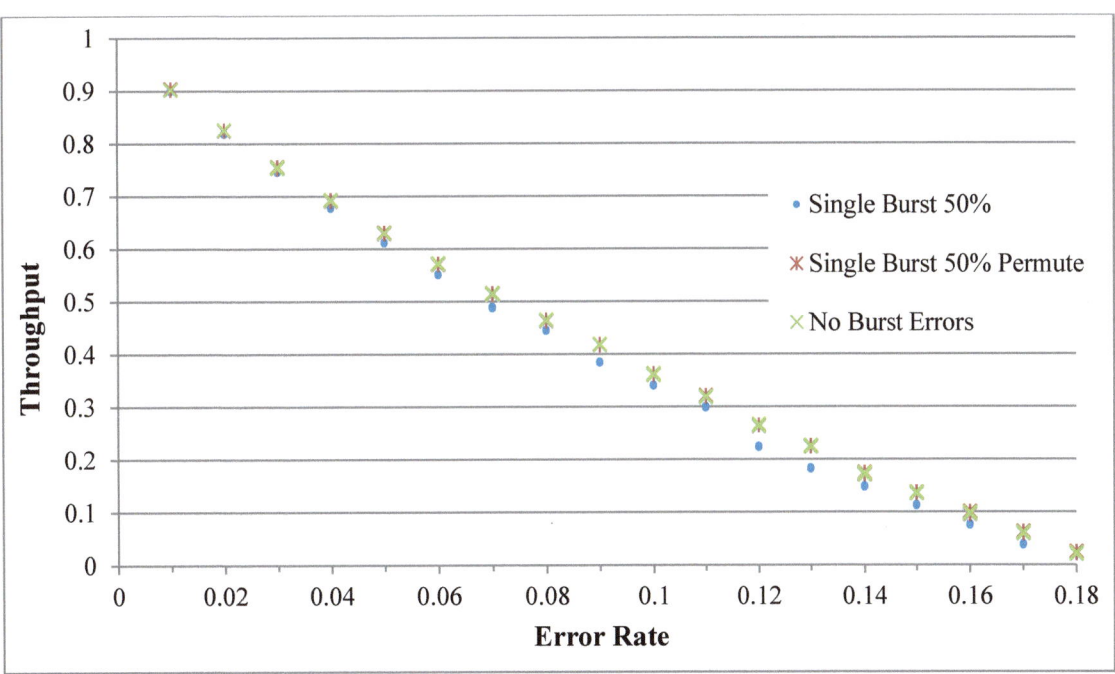

Figure 17. Throughput for a Single Burst Error (50% of Errors contained in Burst)

From the plot, it is clear that single burst errors do have an effect on the throughput of the Cascade protocol. This can be seen especially at higher initial error rates starting from 12% and higher. Applying a random permutation to the sifted key before the initial pass appears to have mitigated the effects of single burst errors as the performance with a permutation is similar to the random error case without burst errors. Essentially, there appears to be no difference between the performance of Cascade with and without burst errors as long as a permutation is applied before the first pass.

Figure 18 shows a single burst error with an error ratio of 100%. The impact of the single burst error is magnified when the burst error contains all the error bits. In fact, the throughput drops by up to 8% at error rates below 12%. The effect at and above 12% error rate is even higher, with an average throughput reduction of up to 15%. At error rates greater than 16%, the throughput drops to zero as all bits are effectively exposed. As with the 50% ratio burst, a permutation of the first pass mitigates the effect of the burst error and the permuted performance approaches that of the random error case. Both the large and small single burst error cases show a smaller effect on throughput at lower error rates. For instance, the throughput for small single bursts is negligibly impacted for error rates of 4% and smaller. Similarly, large single bursts do not being to significantly impact the throughput until around 3% error rates.

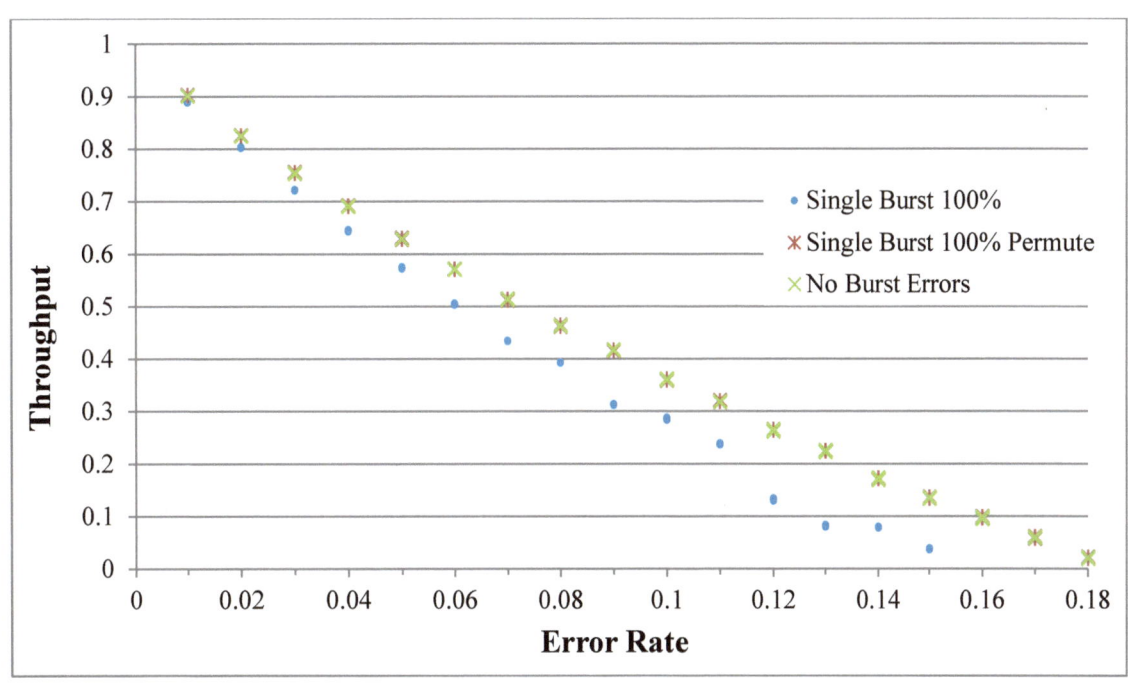

Figure 18. Throughput for a Single Burst Error (100% of Errors contained in Burst)

The plot shown in Figure 19 shows the data collected for a periodic burst error case with 1000 bursts containing 50% of the error bits and initial error rates of 1% to 18%. The x-axis is the chosen error sampling rate, and the y-axis is the estimated error rate. Again, for error rates above 18% the algorithm exposes 100% of the sifted key bits. The plot shows Cascade's throughput for a sifted key with random error distribution, a 1000 burst errors with an error ratio of 50%, and 1000 burst errors with permutation before the first pass.

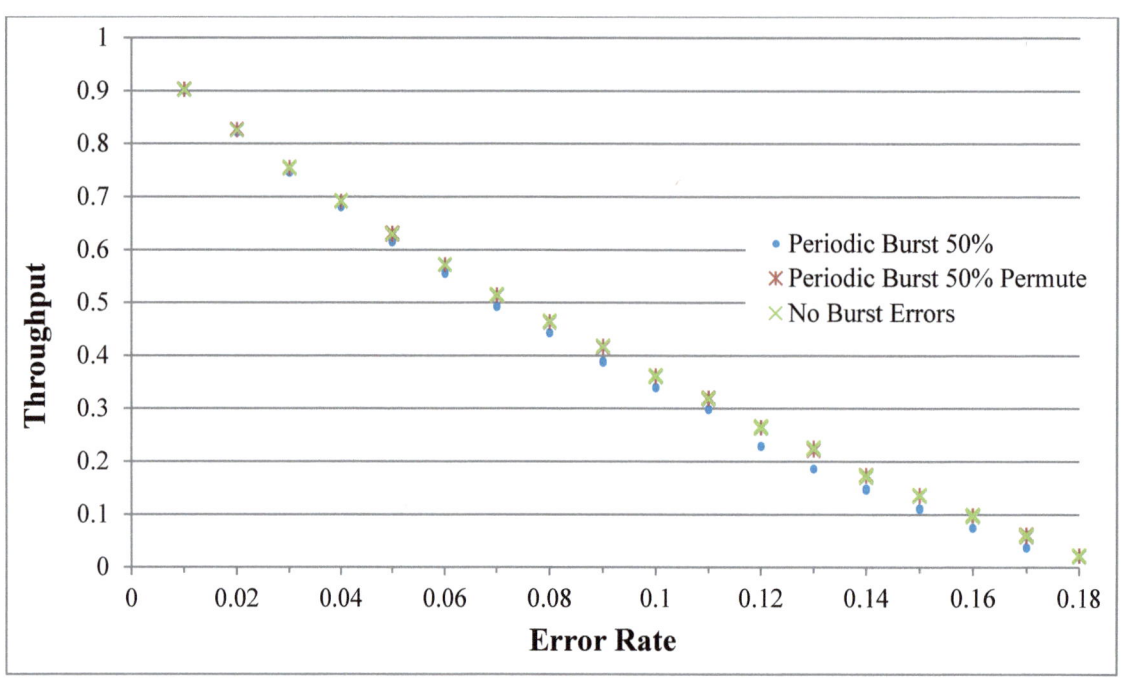

Figure 19. Throughput for 1000 Periodic Burst Errors (50% of Errors contained in Bursts)

It is apparent from the plot that Cascade is affected by periodic burst errors as well. The impact of the periodic burst errors is greater at higher error rates especially starting at 12%. Interestingly, smaller periodic burst errors do not significantly affect the throughput until 5% error rates and the periodic burst error case performed better at all higher error rates than the single large burst error. This is most likely due to the larger proximity between errors and correct bits as compared to one single large burst. As in the other cases, a significant decrease in throughput is evident at 12% error rates and higher. A random permutation of the key before the first pass of Cascade results in throughput that is similar to the simple random error distribution.

Figure 20 shows periodic burst errors with an error ratio of 80% (due to the number of bursts and burst lengths, an error ratio of 100% was not possible for periodic bursts in the simulation). The impact of the periodic burst errors is magnified when the burst error contains nearly all the error bits. The drop in throughput approaches 8% near

an 8% initial error rate and increases as the error rate increases. As with the 50% ratio

burst, a permutation of the first pass mitigated the effect of the burst error and the

permuted performance approaches that of the random error case. Interestingly, for the

non-permuted case, all bits were exposed for error rates over 16%. After the permutation,

error rates above 18% exposed all bits for zero throughput.

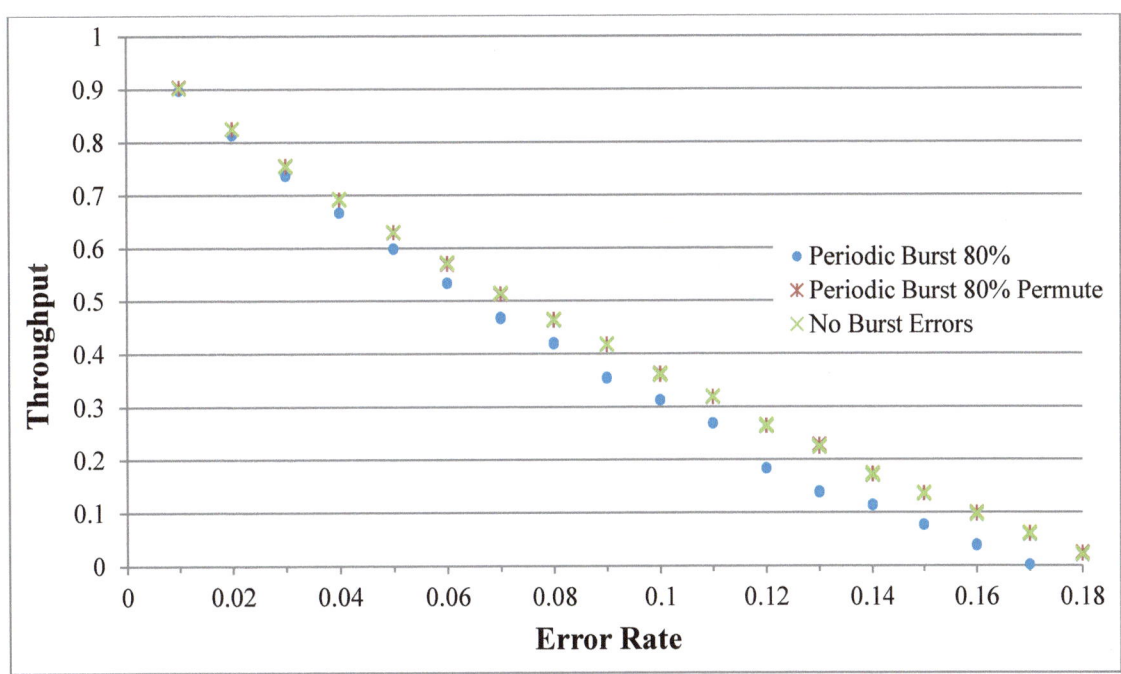

Figure 20. Throughput for 1000 Periodic Burst Errors (80% of Errors contained in Bursts)

Overall, the Cascade protocol is quite resistant to both single and periodic burst

errors up to a 16% initial error rate. However, the throughput can still be significantly

improved for error rates between 4% and 18% by randomly permuting the sifted key

prior to the first pass. The performance of Cascade is also not affected by a permutation

for random error distributions. As a result, by always permuting the sifted key prior to the first pass, the affects of any random or systemic burst errors can be alleviated, maximizing available throughput. Permutation can break up clumps of errors that occur naturally in random error distributions, reducing error masking and raising throughput.

4.6 Cascade Protocol Performance

The throughput of an error detection and correction algorithm is defined as the percentage of the sifted key that remains after reconciliation is performed. The theoretical limit on throughput is given by Kollmitzer and Pivk and the minimum number of bits exposed by any reconciliation protocol is the Shannon Entropy of the error rate as given by Equations 2 and 3, defined above in Section 2 (Kollmitzer & Pivk, 2010) (MacKay, 2005). Throughput is directly affected by the initial error rate and a higher error rate is expected to increase the number of bits exposed due to increased parity checks and subsequent discarding of the exposed bits. As a result, a smaller initial error rate should result in a higher throughput that is closer to the theoretical bound. Figure 21, shown below, reveals the theoretical bound on throughput as the solid line along with the experimentally discovered throughput using randomly distributed errors for Cascade. The throughput is measured as bits remaining after reconciliation divided by total sifted key bits. The plot reveals that Cascade performs exceedingly well at lower initial error rates and the performance drops as error rate increases. The throughput does not seem to be affected greatly by sifted key length. A key length of 10,000 bits varies slightly in throughput compared to 500,000 bits but overall, the difference is not significant. Depending upon the final key size needed for the privacy amplification step, the Cascade algorithm can yield usable key bits up to around 18% error rate. However, in practice an error rate above 15% usually results in the forced failure of the algorithm as an error rate above 15% is seen as evidence that an eavesdropper is present, thus failing to meet the established error rate abort parameter. Sugimoto and Yamazaki and Yan et. al. found similar throughput results as shown in Table 4. However, both used different implementations of Cascade with differing initial block sizes.

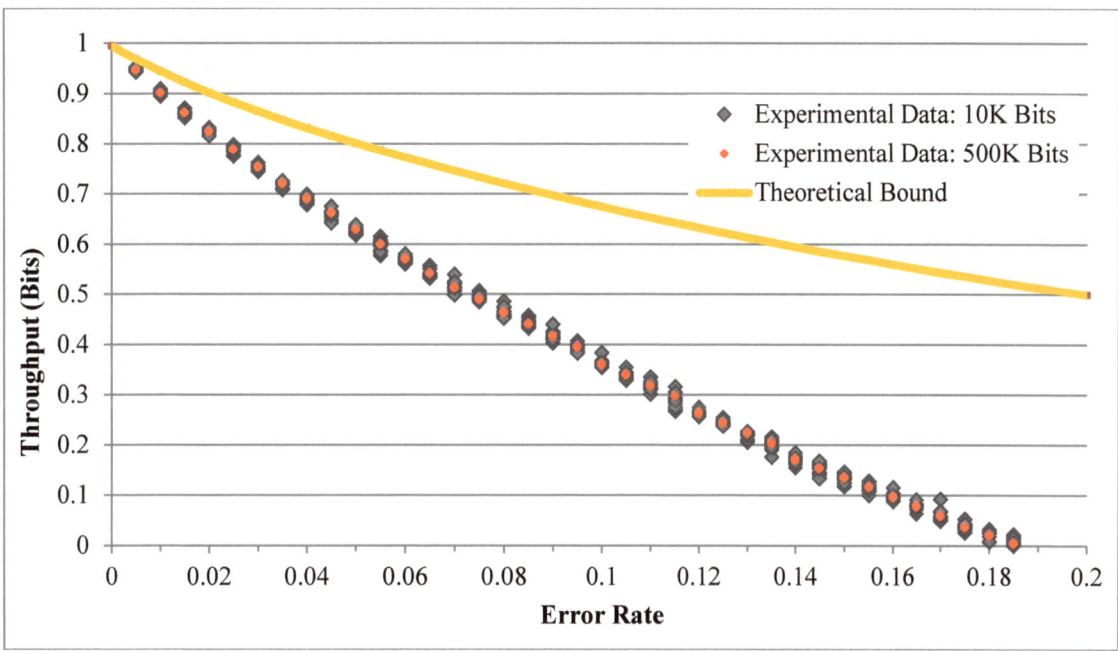

Figure 21. Cascade Theoretical and Experimental Throughput

There does appear to be an interesting difference between the throughputs and error rates for the data presented in this research and the data found by Sugimoto and Yamazaki and Yan et.al. At 1% error rate, there is roughly a 0.5% reduction in throughput versus the literature results. At 3% error rate, the throughput difference is 1%, the difference at 3% error rate is 2%, and the difference at 15% error rate is nearly a 10% reduction in throughput. This effect is most likely due to design choices in protocol implementation and experimental setup. In addition, there does not appear to be an appreciable throughput difference between shorter and longer sifted key lengths.

Initial Error Rate	Author (500K bits, Dynamic blk sz)	Author (10K bits, Dynamic blk sz)	Sug* (10K bits, 73 blk sz)	Yan** (10K bits, 73 blk sz)
0.01	0.9020	0.9017	0.9090	0.9067
0.03	0.7535	0.7531	0.7735	----------
0.05	0.6289	0.6278	0.6609	0.6623
0.075	0.4897	0.4956	0.5333	----------
0.10	0.3605	0.3644	0.4233	0.4139
0.125	0.2443	0.2436	0.3245	----------
0.15	0.1353	0.1346	0.2305	0.2187

* (Sugimoto & Yamazaki, 2000) ** (Yan, Ren, Peng, Lin, Jiang, & Liu, 2008)

Table 4. Cascade Throughput Comparison from Prior Research

Figure 22 gives the bits exposed as a function of error rate for the Cascade algorithm. The resulting line is not quite linear (dotted line), exhibiting more of a polynomial trend. For this graph, the initial key length was 500,000 bits. For a 10% error rate, slightly less than half of the initial key bits have been exposed. At 20% error nearly all of the bits have been exposed during the operation of the algorithm.

Figure 23 shows the theoretical and experimental leakage for the implementation of the Cascade algorithm. This graph is the complement to the throughput graph as it shows the fraction of exposed bits versus the error rate. As with the throughput, the fraction of bits exposed is very low for low error rates and increases as error rate increases. The bit sampling rate used for the Cascade algorithm has a large effect on the experimental leakage and throughput since the bits sampled are discarded before the algorithm is executed. As a result, optimizing the bits sampled becomes a tradeoff between higher throughput or better error estimation and algorithm performance. If error estimation is inaccurate, throughput decreases as more parity checks are needed to correct the errors due to incorrect choice of block size and number of blocks.

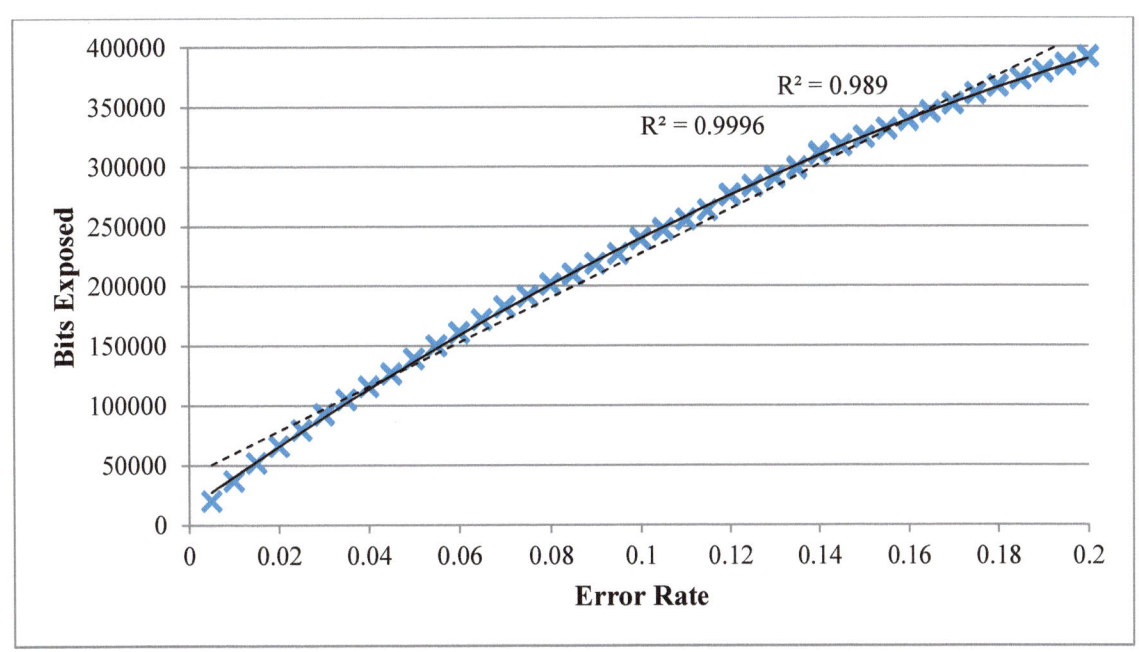

Figure 22. Cascade Bits Exposed as a Function of Error Rate

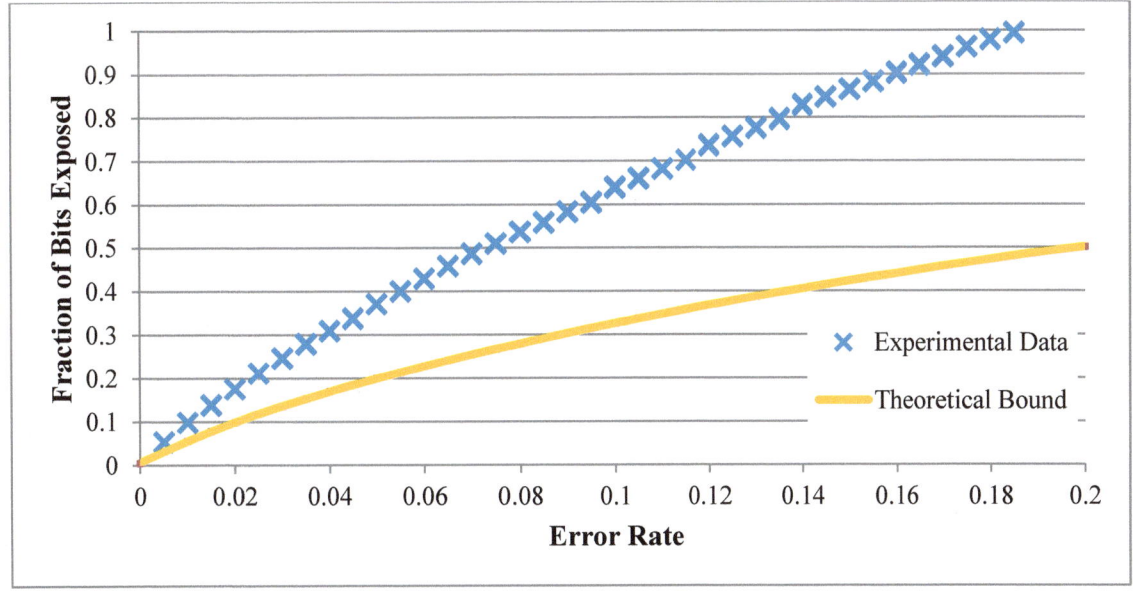

Figure 23. Cascade Theoretical and Experimental Leakage

4.7 Summary

This chapter presented the results of the experiments described in Chapter 3 and discussed conclusions drawn from the data. The following chapter will discuss the overall results of this research and make recommendations for future research.

V. Conclusions

5.1 Research Summary

Quantum key distribution is an interesting area within the field of cryptography. It is only the latest step in the long history of methods used to ensure that secret information remains secret. Unfortunately, even though the dawn of QKD began in the early 70's and 80s, our technological progress and level of achievement has not kept pace with the ideas and theories regarding quantum mechanics, although it is hastening the obsolescence of classical encryption methods and schemes rapidly. Although the theory that a true ideal quantum key distribution system is perfectly secure is supported, the technology needed to create an ideal quantum key distribution system is out of reach for the time being and finding solutions for the problems on the horizon are important. But the technology that is available is good enough to implement quantum key distribution using less than ideal components with the addition of several important steps to minimize weaknesses in hardware and implementation. The steps that make quantum key distribution possible using today's technology are employed within various protocols and implementations that add steps such as error detection and correction and privacy amplification to imperfect hardware. As a result, powerful quantum key distribution systems are being researched, studied, and even used in real world transactions.

Reconciliation protocols are used in every quantum key distribution realization as a result of the importance of ensuring the detection of simple transmission errors and revealing if an eavesdropper is present. Due to real world limitations in transmission and detection systems as well as potentially lossy transmission media and eavesdroppers, reconciliation protocols are critical to the operation of QKD systems using current technology. The Cascade secret key protocol is one of several that is used for reconciliation and although there has been some study into the efficacy and efficiency of the protocol, much of the research is focused on aspects of the algorithm's operation using inputs that tend to be focused in a very confined problem space such as small key lengths (faster processing), while others, such as (Maurer, 1993), and (Lo, 2003), only delve into mathematical proofs and theory. The purpose of this research was to explore the limits of the protocol and attempt to find ideal or more ideal parameters that could be used to optimize the performance of Cascade. Factors such as error sampling rate, key length, throughput, information leakage, and the protocol's performance in the presence of various error distributions were explored in an attempt to expand the knowledge of the workings of Cascade and improve upon its performance.

5.2 Findings

The error sampling rate is an important factor of Cascade in that it is ultimately used with the desired errors per block in the calculations for block size and number of blocks in the form of an estimated error rate. As a result, the error sampling rate and subsequent error estimation have a direct effect on the accuracy and efficiency of Cascade for error correction. In the research, it was found that the error sampling rate does not adversely affect bit leakage, but is an important factor in error estimation accuracy. As the error sampling rate is reduced, the estimated error rate begins to deviate from the actual error rate by up to 3.5 times when 5% error sampling rate is chosen over 10% error sampling. Conversely, the higher the error sampling rate, the more likely the

61

estimated error rate is to accurately reflect the actual error rate. In addition it was demonstrated that the chosen sampling rate does not impact the final bits exposed in the corrected key used for privacy amplification. The potential benefit of choosing ideal error sampling rates becomes negligible after about 25%. It was shown that the commonly used error sampling rate of 50% is inefficient and wasteful and an error sampling rate of 25% is suggested to maintain a balance between error rate estimation and usable sifted key bits.

Error estimation and sifted key size are critical to the efficient operation of Cascade as well. The Cascade protocol was shown to be very accurate in error rate estimation up to 18% initial error rate with average error rate deviation of less than 5% for 1% initial error rates and less than 1% for error rates up to 20%. As a result, the protocol is able to correctly calculate block size and number of blocks sufficient to detect and resolve errors efficiently. The research revealed that that the deviation between actual error rate and estimated error rate increases as sifted key length decreases with the largest increase in error estimation deviation occurring in the smallest initial error rate tested, 1%. The data show that shorter sifted keys result in decreased error estimation power especially at lower initial error rates. As a result, the Cascade protocol can tend to over or underestimate the number of blocks and block size during the Binary bisection, resulting in more passes required to detect and correct error bits. However, for sifted keys of 750,000 bits or greater, the improvement in error estimation is negligible and it is unlikely that improving error estimation deviation to less than 1% will have any additional benefit to the performance of Cascade. An interesting finding is that the sifted key length has a direct impact on the number of bits exposed or leaked from the protocol.

The maximum ratio of bits exposed is reduced as key length increases, although the minimum percentage of bits exposed increases as key length increases. The data show an asymptotic trend of maximum and minimum bits exposed leveling out as the sifted key length increases. As a result, it appears that by choosing a key size within the "sweet zone" between ideal minimum and maximum exposure, one can optimize for maximum available key remaining for privacy amplification and retain the best possible ratio of usable bits to discarded bits. The most important takeaway regarding sifted key length is that the length of the input key chosen does not just affect the size of the reconciled key; it actually has a direct impact on how well the Cascade protocol operates. In order to optimize QKD as much as possible, thought should be given to balancing the minimum acceptable reconciled key length with the effect that that the requisite initial sifted key will have on the efficiency of the reconciliation protocol when Cascade is utilized.

Cascade was shown to perform fairly well on several error distribution cases involving single and periodic burst errors, but it was found that a permutation before pass zero restored the throughput back to random error distribution performance. Cascade was found to be more robust against burst errors than initially expected, although burst errors do have a negative effect on the throughput of the protocol especially at higher initial error rates. In fact, it was found that single burst errors induce more degradation to throughput at higher error rates than periodic burst errors. However, applying a random permutation to the sifted key before the initial pass mitigates the effects of single burst errors. This is a result of the error separating effect that the permutation forces, essentially reorganizing the burst error bits into a distribution that is more like a random burst-free distribution. A modification to Cascade to always permute the sifted key prior to the first pass is recommended to minimize the effects of potential burst errors and allow the protocol to operate at maximum throughput.

Overall, Cascade performs exceedingly well at low error rates through 10% but is still robust enough to yield usable key bits up to 18% error rate if required. By carefully tweaking the input parameters while understanding that they are interrelated, one can effectively maximize the operational efficiency and usefulness of the Cascade secret key reconciliation protocol and ensure that it continues to provide accurate and reliable error detection, correction, and eavesdropper awareness for today's quantum key distribution systems.

5.3 Limitations

There are several important limitations to the research presented in this paper. First, the implementation of the Cascade protocol was intended to be as close as possible to the original proposed by Brassard and Salvail in 1994. Some improvements made by Brassard and Salvail to the protocol were included in this implementation, specifically, logic for how many passes are run before the protocol finishes. Another limitation is the method utilized to provide random generation of the sifted key and the initial introduced errors. The code utilizes the Mersenne Twister pseudo-random number generator. A cryptographically secure pseudo-random number generator may be a better choice for future implementations of Cascade simulation software and increase the fidelity of the operation as compared to actual hardware implementations of the protocol in quantum key distribution. A final limitation of the research presented in this paper is that the software is monolithic in nature. In other words, the simulation of both Alice and Bob's buffers and Cascade interactions were conducted using only one software implementation. In reality, the only information that Alice and Bob have in common is the information passed over the quantum and public channels and processing of their respective sifted keys is done separately by each party.

5.4 Future Research Recommendations

There are several areas of improvement that could be performed in future research in order to advance the operational efficiency of the Cascade secret key reconciliation protocol. The following is a list of areas that the Cascade protocol would benefit from further study.

1. Along with Cascade, there are many other protocols and algorithms that are used for error detection and correction. Some of these protocols include Winnow, Turbo Codes, and Low Density Parity Check codes. In addition, there is a wide variety of variation in the implementations of the algorithms with many different instantiations and modifications of each. A thorough comparison of the Cascade protocol with other error detection and correction protocols and their implementations would be critical in determining which protocols work best in different situations.

2. Improvements and optimizations to the C++ implementation of the Cascade Protocol used for this thesis could increase both the fidelity and simulation efficiency of the software. Additional algorithm implementations could be added to the software to compare different tweaks and modifications to the Cascade protocol.

3. An in depth study of the permutations performed in the Cascade Protocol could greatly improve upon the efficiency of the protocol in distributing error bits within the blocks. The distribution of error bits during the various passes of the protocol depends upon the permutation chosen. It is likely that some permutations are better than others in breaking up blocks of errors that might increase the number of parity checks needed to correct errors, resulting in reduced throughput and more information leakage. By finding optimal permutations, leakage could be reduced and throughput increased resulting in an overall improvement in the efficiency of the Cascade protocol.

4. A deeper view into information leakage during the public channel communications between Alice and Bob during Cascade could also be performed in order to further characterize exactly how much information is being leaked. Currently, error detection and correction protocols treat every leaked bit as having revealed the same amount of information to a potential eavesdropper, in essence assuming the worst case scenario. If certain bit transactions reveal more information than others, it would make sense to characterize the amount of information leaked and adjust the privacy amplification to optimize information leaked and improve overall throughput by only discarding bits in transactions with higher probability of leakage while retaining bits with low probability. This would be especially beneficial in situations where sifted key lengths are constrained and every bit is needed to ensure operation of quantum key distribution in priority environments.

5. Although software based simulations are useful for exploring the problem space while taking up a relatively light footprint, a more true to reality simulation would include hardware in the loop capability. It would be very beneficial to study the nuances of the Cascade protocol using a hardware implementation of a quantum key distribution setup using FPGA boards and actual hardware to pull out finer details about the operation of the protocol within a realistic environment. With software simulations, only the essence of reality can be captured and if an effect of behavior is missed and not encoded, it will not affect the simulation. Since most commercial quantum key distribution system protocols are "hardwired" and unchangeable, a fully parameterized hardware implementation of Cascade and other protocols within a realistic QKD system would allow for the true trade space of the algorithms to be explored, increasing the likelihood of true optimizations that could drastically improve the efficacy and efficiency of working QKD systems.

Appendix A: BB84 QKD Protocol Example

Quantum Channel

Alice's Random key Bits	1	0	0	1	1	1	0	0	0	0	1	1	0	1	0	0	0	1	1	0
Alice's Random basis	+	+	X	+	X	X	+	X	X	+	+	+	X	+	+	X	X	+	X	+
Alice's qubit polarization	⇔	⇔	⤢	⇕	⤡	⤢	⇔	⤢	⤡	⇕	⇔	⇕	⤢	⇕	⇔	⤡	⤡	⇕	⤢	⇔
Bob's random basis measurement	+	+	+	+	X	X	+	+	X	+	+	X	X	+	+	X	X	X	X	+
Bob's received bits	1	-	1	1	0	1	0	1	-	0	1	1	0	1	0	-	0	1	-	0

Public Channel

Bob announces basis choices	+		+	+	X	X	+	+		+	+	X	X	+	+		X	X		+
Alice sends basis confirmation	☑		☑	☑	☑	☑	☑	☑		☑	☑	☑	☑	☑	☑		☑	☑		☑
Alice and Bob share sifted key	1		1	1	1	0				0	1		0	1	0		0			0
Bob reveals random bit subset	1			1						0										0
Alice confirms bits	☑			☑						☑										☑
Alice and Bob share secret key			1		1	0				0	1			1	0		0			

+	Rectilinear Basis
X	Diagonal Basis
⇔	0-degree Polarization
⇕	90-degree Polarization
⤡	135-degree Polarization
⤢	45-degree Polarization
-	No bit received
☑	Successful check (matching basis or bit)

Appendix B: Cascade Protocol Pseudo Code

```
Binary Pseudo Code

doBinary(Pass, Min, Max, InCascade)
{
  Validate input parameters are legal
  Determine Range for this block
  Increment bits exposed
  Calculate parity of Alice(Min,Max)
  Calculate parity of Bob(Min,Max)
  Compare parity bits
  if(Parity Does Not Match)
  {
    switch(Range)
    {
      case 0:
        // Single bit range
        Correct Error
        if( (Pass > 0) && (Not InCascade) )
        {
          doCascade(Pass, Min);
        }
        break;

      case 1:
        // Two bits range, so check the first one
        Calculate parity of Alice(Min,Min)
        Calculate parity of Bob(Min,Min)
        Increment bits exposed
        Compare parity bits
        if(Parity Does Not Match)
        {
          // first bit mismatches
          Correct Error
          if( (Pass > 0) && (Not InCascade) )
          {
           doCascade(Pass, Min);
          }
        }
        else
        {
          // second bit mismatches
          Correct Error
          if( (Pass > 0) && (Not InCascade) )
          {
           doCascade(Pass, Max);
          }

        }
        break;

      default:
        // Range is more than 2 bits
        // check first half
```

```
            low1 = Min;
            high1 = Min + ( Range / 2 );
            low2 = high1 + 1;
            high2 = Max;
            Calculate parity of Alice(low1,high1)
            Calculate parity of Bob(low2,high2)
            Compare parity bits
            if(Parity Does Not Match)
            {
                // error in the first half
                doBinary(Pass, low1, high1,0);
            }
            else
            {
                // error in the second half
                doBinary(Pass,low2,high2,0);
            }
            break;
        }
    }
}


doCascade(Pass, ErrorBit)
{
    for(Each pass from 0 to Pass)
    {
        Find block min and max for this pass
        Find Index for this block this pass
        doBinary(Index,Min,Max,1);
    }
}


SimulateCascade()
{
    Start timer
    Get command line arguments
    Open output files
    Initialize variables
    Initialize Permute and Reverse Permute arrays
    Initialize Random Number Generator

    For(each trial from 0 to NumberOfTrials)
    {
        Initialize all bit buffers

        if(UseAliceBobFile)
        {
            Read input from files
            Identify errors between Alice and Bob
            Calculate actual error rate
        }
        else
        {
            Fill Alice's buffer with random data
            Copy Alice's buffer to Bob's buffer
```

```
        Corrupt Bob's buffer with errors based upon command line
selection
        Permute buffers, if necessary
        if(Use Sampling)
        {
          Determine estimated error rate
        }
        else
        {
          Use user supplied error estimation
        }
    }
    Determine initial block size
    Determine last pass

    for(each Pass)
    {
        for(each Block)
        {
            doBinary(CurrentPass, low, high, 0);
        }
    }
    Calculate statistics
    Write statistics
}
```

Bibliography

Ardehali, M., Chau, H. F., & Lo, H.-K. (1999, Jan 29). *arXiv:quant-ph/9803007v4*. Retrieved Mar 4, 2011

Bennett, C. H., & Brassard, G. (1984). Quantum Cryptography: Public Key Distribution and Coin Tossing. *International Conference on Computers, Systems & Signals Processing*. Bangalore, India.

Bennett, C. H., Bessette, F., Brassard, G., Salvail, L., & Smolin, J. (1992). Experimental quantum cryptography. *Journal of Cryptology*, *5* (1), 3-28.

Boughattas, M. B., Iyed, B. S., & Rezig, H. (2010). Correcting codes in the quantum keys reconcilation: scenarios of privacy maintenance. *IEEE International Conference on Privacy, Security, Risk and Trust*, (pp. 1022-1025).

Brassard, G. (1993). A bibliography of quantum cryptography. *ACM SIGACT News*, *24* (3), 16-20.

Brassard, G., & Salvail, L. (1994). Secret-key reconciliation by public discussion. *Workshop on the theory and application of cryptographic techniques on Advances in cryptology* (pp. 410-423). Lofthus, Norway: Springer-Verlag.

Capraro, I., & Occhipinti, T. (2007). Implementation of a Real Time High Level Protocol Software for Quantum Key Distribution. *IEEE International Conference on Signal Processing and Communications*, (pp. 704-707).

Chen, K. (2000, Aug). Improvement of Reconciliation for Quantum Key Distribution. *Master's Thesis*. Rochester Institute of Technology.

Ekert, A. (1991). Quantum cryptography based on Bell's theorem. *Phys. Rev. Lett.*, *67* (6), 661-663.

Ferguson, N., & Schneier, B. (2003). *Practical Cryptography*. Indianapolis, Indiana: Wiley Publishing, Inc.

Kollmitzer, C., & Pivk, M. (Eds.). (2010). Applied Quantum Cryptography. *Lecture Notes in Physics*, *797*, *1*. Springer, Berlin Heidelberg.

Lo, H. K. (2003). Method for decoupling error correction from privacy amplification. *New Journal of Physics*, *5*, 36.1-36.24.

MacKay, D. J. (2005). *Information Theory, Inference, and Learning Algorithms* (4th Edition ed.). Cambridge: Cambridge University Press.

Matsumoto, M., & Nishimura, T. (1998). Mersenne twister: a 623-dimensionally equidistributed uniform pseudo-random number generator. *ACM Trans. Model. Comput. Simul.*, *8* (1), 3-30.

Maurer, U. M. (1993). Secret key agreement by public discussion from common information. *IEEE Transactions on Information Theory*, 733-742.

Nakassis, A., Bienfang, J., & Williams, C. (2004). Expeditious reconciliation for practical quantum key distribution. *Quantum Information and Computation, Proc. SPIE 5436.*

Nguyen, K.-C. (2002). Extension des Protocoles de Réconciliation en Cryptographie Quantique. *Masters thesis*. Université Libre de Bruxelles.

Pearson, D. (2004). High-speed QKD reconciliation using forward error correction. *Proceedings of the 7th International conference on Quantum Communication, Measurement and Computing (QCMC)*, (p. 299).

Rass, S., & Kollmitzer, C. (2009). Adaptive error correction with dynamic initial block size in quantum cryptographic key distribution protocols. *Third International Conference on Quantum, Nano and Micro Technologies*, (pp. 90-95).

Scarani, V., Acín, A., Ribordy, G., & Gisin, N. (2004). Quantum cryptography protocols robust against photon number splitting attacks for weak laser pulses implementations. *Phys. Rev. Lett.*, *92* (5), 057901-1-057901-4.

Schneier, B. (1996). *Applied Cryptography: Protocols. Algorithms, and Source Code in C.* (2nd Edition ed.). John Wiley & Sons, Inc.

Serna, E. E. (2009, August 19). *arXiv:quant-ph/0908.2146v3*. Retrieved April 20, 2011, from http://arxiv.org/PS_cache/arxiv/pdf/0908/0908.2146v3.pdf

Singh, S. (2000). *The Code Book: The Science of Secrecy from Ancient Egypt to Quantum Cryptography.* Anchor Books.

Sugimoto, T., & Yamazaki, K. (2000). A study on secret key reconciliation protocol "cascade". *IEICE Trans. Fundamentals*, *E83-A* (10), 1987-1991.

Townsend, J. S. (2000). *A modern approach to quantum mechanics.* Sausalito, California: University Science Books.

Van Assche, G. (2006). *Quantum Cryptography and Secret-Key Distillation.* Cambridge: Cambridge University Press.

Wiesner, S. (1983, Winter-Spring). Conjugate Coding. *ACM SIGACT News - A Special Issue on Cryptography*, *15* (1), pp. 78-80.

Yamazaki, K., Nair, R., & Yuen, H. P. (2006). Problems of the CASCADE Protocol and Renyi Entropy Reductions in Classical and Quantum Key Generation. *QCMC*. arXiv:quant-ph/0703012.

Yan, H., Ren, T., Peng, X., Lin, X., Jiang, W., & Liu, T. (2008). Information Reconciliation Protocol in Quantum Key Distribution System. *Fourth International Conference on Natural Computation*, (pp. 637-641). Jinan.